Traffic Life

Traffic Life

Passionate Tales and Exit Strategies

Edited by Stephan Wehner

An Anthology

wandering soliton publications
2004

Cover Design: Andrea Schmidt (www.a-schmidt.com)

This work contains texts from around the world. In the absence of a standard for English spelling, the authors' choices were used.

99.9% free of car brand names. (The references in Harlan Ellison's story *Along the Scenic Route*, written in the 1960's, were not removed)

National Library of Canada Cataloguing in Publication Data

Traffic life: passionate tales and exit strategies /editor: Stephan Wehner.

Includes bibliographical references.

ISBN 0-9734022-0-2

1. Automobiles—Social Aspects—Literary collections. 2. City traffic—Literary collections. 3. Automobiles in literature. 4. Automobiles in art. 5. Automobiles—Social Aspects—Songs and music. I. Wehner Stephan, 1966–

NX650.A86T73 2003 700'.455 C2003-906667-3

First Edition.

Printed in Canada.

Text paper stock is 100% Recycled Post-Consumer Waste and Process Chlorine-Free ("PCF"). It is not necessarily "Totally Chlorine-Free", since chlorine or chlorine derivatives may have been inserted into the fibre for an earlier paper life.

wandering soliton publications
(a division of)
The Buckmaster Institute, Inc.
2150 Adanac Street
Vancouver British Columbia V5L 2E7
Canada

Contents

Foreword

Transportation professionals often talk about the four "E's" of traffic—engineering, education, enforcement, and encouragement. All four aspects are needed to have a transportation system that works effectively. But what happens when our infrastructure is skewed towards the driver at the expense of the pedestrian? What happens when drivers never learn the crosswalk laws? What happens when driving laws are seldom enforced? And what happens when pedestrians are only encouraged to get out of the way? "E" could be the grade we give the resulting environment.

Put a kind and considerate human being behind the wheel of an automobile in this environment, and an ugly transformation takes place. The kindergarten wisdom of taking turns and following rules goes out the tailpipe. The horn is preferred to the brake pedal. People in other cars are competitors for your space. People outside of cars are an obstruction and a nuisance. Just as importantly, people who don't drive become less worthy.

Americans are all too familiar with this driving culture. The current generations grew up immersed in it. Cruising the strip was the central part of life in "American Graffiti", and in a thousand other movies. It is repeated every hour of every day in thirty second increments, thanks to the billion dollar advertising budgets of the automobile industry.

Suppose we could break free from the windshield perspective. Imagine for a minute "American Graffiti" with the teenagers on Schwinns. Imagine them strolling on foot along the Ramblas in Barcelona, or between sidewalk cafes in Paris. Imagine the blonde in the sports car riding the Metro. And imagine all that in an American suburb.

However, we are not yet free to be walkers in a walking world. We are currently pedestrians and bicyclists struggling to exist in an autocentric world. A world that is indifferent—and hostile—to us. A world of traffic.

As easy as it is to grow cynical, deep down inside I am inherently an optimist. Most advocates are. We would never work so hard for so little financial reward if we didn't believe that this world could be a better place. But where does one start?

The public needs to realize that there is a problem, and that there are solutions that do not involve more cars and more roads. That is why I produce my cable television series, *Perils For Pedestrians.* Raising public awareness of the issues is an important step towards fomenting change.

So it is with the arts. Good poems, stories, drawings and music capture the essence of a moment. They can draw out a hidden meaning. They can inspire a vision.

A television program, a book, or a website will never replace the grass roots advocates testifying at endless planning board meetings. But they may inspire someone to go to that meeting, and their ideas might be received just a little bit more warmly.

I hope that you, the reader of this collection, will gain a vision of what could be. If you help spread that vision and act on that vision, it will only be a matter of time before we see the necessary policy changes.

Someday we may finally tame the most defining characteristic of post World War II suburban development—Traffic.

John Z. Wetmore
Bethesda, Maryland
December 2003
Producer of "Perils for Pedestrians"
www.pedestrians.org

Preface

When I was living in San Francisco, I would go to one of the coffee shops to enjoy my coffee before cycling to work. While I sat at a sidewalk table, absorbing the morning, I would hear over and over again, "one large-soy-decaf-latte-to-go" and would watch as this latte was carried to the street, to be swallowed into the anonymous interior of a vehicle. To this day, I can't see the point of drinking coffee in a car.

While moving from San Francisco, California, to Vancouver, British Columbia, I was sitting in traffic in a rental truck. I looked around and saw car after car in front, behind, beside, and sometimes over and under me, and I thought: how silly this is. Here we are, all moving mechanically—somewhere. Bodies stationary in a moving frame. Exchanging places. One should have a book of nice little stories about this. A book that covers values, feelings and addresses the cultural ethos.

We have a car culture that has produced communities of thousands of people without a store, where everyone arrives home and leaves again by car, where neighbours hardly know each other, isolated by an insidious mechanism.

A mechanism that is fed by exaggerations and misconstructions. Look at ads for cars: Grandiose, glamorous, brilliant. Landscapes, open roads, speed, agility and excitement flash and bleat at us from every media outlet. Yet everyone knows the banality. Reality is different. All but a few cars spend their time just sitting around—or in traffic. And cars, cars in great numbers, produce traffic. Most driving is boring. Roads are often packed with cars. City streets get clogged.

When I was still in San Francisco, my employer wisely offered to cover a monthly public transportation pass for all staff. When the possibility of relocation of the company came up, I feared the move, but he said that we would only move within walking distance of a BART station (BART is one of the local transit organizations). In contrast to this, one of my colleagues, a follower of the car religion, rented a house in the suburbs which was too far from the BART station to walk, and so he drove there every morning, to spend half an hour on the train. On one of his trips his car was hit by another car. Beyond the injuries, he had to deal with medical and insurance issues. Another friend became unreasonably angry while driving. He regularly underwent the "ugly transformation" described by Mr. Wetmore in the foreword to this book. All this was going through my head when I thought "How silly this is!"

How did this collection come about? After arriving in Vancouver, with only a vague idea of what could come out of this request, I distributed a call for contributions. How would others express these problems or offer solutions? Contributions arrived, and some of them surprisingly fit the picture that had been forming in my head—some made me feel, "oh, that's what I was thinking of!" Others, I couldn't relate to at first—until I realized, yes, they also talk about the reality, the pain, the suffering, the stupidity, the ultimate silliness.

Not every piece arrived through the call for contributions. In addition to some that I came across, I also added some special items: the manifesto of the New York organization, Right of Way; *Declaration of the Right to Walk and Roll* by the group WALK Austin; and the *Autoholics Anonymous Car-Free Pledge* developed by Carbusters. I found a very nice book by Aaron Naparstek, *Honku: The Zen Antidote to Road Rage*, and am glad that he agreed to contribute an excerpt. Jeff Younger composed a Jazz piece for this book. The scores are included (if you can't play it yourself, hopefully you can locate someone who will). Finally, I added two widely known short stories: Ray Bradbury's *The Pedestrian* and Harlan Ellison's *Along the Scenic Route*. Special thanks go to Mr. Ellison for writing a postscript to his story for this book.

For their help in various ways, many, many thanks also go to my partner Tamara Dakic, Albrecht Meyer, Anand Yethiraj, Annalee Heit, David Low, Erick Wong, Helga Low, Laura Quilici, Lisa Lajeunesse, Lisa Miller, Margaret Martin, Paige Raibmon, Taliesin Smith, and Yee Doris Jim.

Stephan Wehner
Vancouver, British Columbia
December, 2003

Touched by an Angel? Nope—Whacked by a Hatchback

Greg Taylor

Frankly, you just don't ever know who you will run into while flying over the hood of a small imported car. Really. One minute you're out riding along on your bike, enjoying a nice sunny Saturday afternoon without a care in the world, and the next you're sailing through space while having a conversation with a guy who introduces himself to you as God.

"Hey, Greg... Gregory! Hi ho! God here. I need to bend your ear for a minute."

"What? Who's there? What's happening?"

"It's Me. God. You know, the Creator of the Universe. The Celestial Chief Executive. I need to talk to you. Got a minute?"

"Wha?"

"A little sluggish on the uptake today, aren't we? OK, I'll speak slowly; try to follow along here. I'm God, and right now you're in the middle of having an accident on your bike. And I'd like to talk with you..."

"God? An accident!?

"Stay with me here, Gregory. A couple of seconds ago you were riding through an intersection on your bike when a 16-year-old kid driving a small hatchback blew through a stop sign right in front of you. You didn't have time to avoid it so you've nailed the side of the car. Hard. Right now you are in the process of flying over the hood. Momentarily you are going to land on the other side of the car in the middle of the intersection. Because your perception of time slows during an accident, this unscheduled trip over the handlebars will seem to take an eternity. So, seeing as you have some time to kill, I figured this would be a good opportunity for us to have a little chat."

"Wait! My bike! Is my bike okay!? Did that idiot hurt my bike?"

"Your bike? Your BIKE!? Gregory, you're poised on the edge of oblivion, actually in the process of (ahem) Meeting your Maker, and you are worried about bits of steel and carbon fiber?"

"Well God, it is a brand new bike, and I'd sure hate to see it get dinged..."

"What the heck is it with you cyclists? Run these guys over with a car, give them a little face-time with the Creator of the Universe to talk things over, and what do I get? Do they want Me to explain the meaning of life? Are they curious about Jesus or Buddha? Heck, I'd be satisfied with a question about whether Elvis is still alive. No! It's the same stupid question every single time! The only thing that these yahoos want to know is whether their precious bike is okay...Gregory! Don't distract me with questions. I need to speak with you right now."

"Uh, sorry God. I didn't know that the 'bike' question was a pet peeve of yours. Hey! Wait a minute! If I'm talking with you, God, does that mean that I'm dead?"

"No, you are not dead. You are going to be a little banged up when this is over, but you most assuredly are not dead. Credit for that small miracle goes to your bike helmet. Besides, I've decided to save something far more entertaining for your Big Exit than getting run over by a car."

"Well, that's a relief, I guess."

"Not to give too much away, but your Grand Finale involves a troupe of female gymnasts, lots of Jell-O, and a poorly wired hot tub. How does this headline sound: 'Septuagenarian Love Shocker: Gal Gymnasts Jolted In Geriatric Jiggle Fest.' You will become an urban legend on your way out. Cool, huh?"

"Wow... really?"

"Hah! Gotcha! I was just kidding with the bit about the gymnasts. You might want to be careful around Jell-O, though. Frankly, not enough people are. Anyway, Gregory, I want to talk with you."

"About what, God?"

"Well, it's about this accident that you're having. I want your little aerial trip across the asphalt this afternoon to serve as a wake-up call. Think of it as an invitation from Me, the Creator of the Universe, to do a priority check on what's going on in your life."

"A wake-up call? Say, wait a minute... is this about me buzzing past that big pack of joggers in the park, God? Man, I'm sorry about that, but those guys are the spawn of Satan. They take up the entire bike path and then just dare you to hit them. Those skinny little assholes have to be taken down a peg or two, dammit, and... oops. Sorry about using the a-word there, God. Just a little nervous, I guess."

"No offense taken, Gregory. I do know what you mean about the runners, though. Definitely not my best work. Actually, it's the dog-walkers who don't pick up after their pets that Satan claims as his own, but I digress."

"Okay, so where am I screwing up, God."

"You're not screwing up, Gregory. At least not yet, anyway. What I want you to do after you hit the pavement is to take stock of what's important in your life, especially the people who love and care for you. You are going to find out today that there are a lot of them, a whole lot more than you suspect."

"Really?"

"Yes, really. Complete strangers will treat you with kindness. Your friends and family will rally around to make sure that you are okay. Oh, and the bicycle thing will get sorted out too, thanks to some very nice people that will be entering your life shortly."

"This sounds great, God. But how is the fact that all of these people will be treating me wonderfully supposed to give me a wake-up call?"

"Man, I must have had you hit your head a lot harder than I realized. The fact that these people will go out of their way to treat you so nicely when you need help should leave you a bit humbled by it all. So what I want you to figure out are ways of living your life that will return that kindness and serve the people around you: Family, friends, strangers."

"That's it?"

"That's it."

"That's too easy."

"Just try it, buck-o. If it was that easy, we could have avoided stuff like the Spanish Inquisition and disco. Especially disco. The seventies were an ugly decade, Gregory."

"So I guess that means that I should go easy on the kid driving the car that just hit me? You know, turn the other cheek and all?"

"I said be nice, not be an idiot. The kindest thing that you can do for this kid is to scare the pants off him. Being kind doesn't mean being a pushover. He nearly killed you through his inattention. That's inexcusable. It will make him a better driver in the future if you make this incident especially memorable. Feel free to go and tear him a new one."

"Okay, got it."

"Well, Gregory, I have to be going."

"Gotcha, God."

"Oh, I nearly forgot. Keep riding your bike. I want you to keep it up, even after this accident. You're at your sanest when you are on your bicycle, notwithstanding the fact that, right now, you appear to be having a conversation with a disembodied voice in your head."

"Thanks God, will do."

"Ciao my son, and Happy Landings."

This story is dedicated to all riders who had their conversation with God turn out differently.

Frogspawn Man versus the Boy Racers

Attila the Stockbroker

Mid nineties.
March.
West Sussex.
I've been to a stream
next to the A27
looking for frogspawn
to populate our brand new garden pond.
I guess I first went there when I was about seven
and have been many times since.
The road is much wider now
the cars are faster
and most of the stream is gone
but one stubborn bit remains
next to the concrete and the cars
and the frogs have obviously had
an orgy of Bacchanalian proportions.
I've found lots of spawn, very quickly.
A glutinous, black-specked mess
fills my bucket.
It's a beautiful spring day.
I'm very happy

full of memories of my father
seven years old again.
I stand by the side of the road
next to the traffic lights
and wait for my lift home.

Suddenly I realise
that an inarticulate-sounding man in his mid twenties
in some kind of penis extension car
has wound down his window
and is shouting abuse at me.
The lights change—
the glans glides off.
Then another man makes a two fingered gesture at me.
A car full of techno nerds
turns down the techno
and hurls a collective techno insult.
The next time the light goes red
a middle aged, middle class
Southwick zip up jumper husband
in a middle aged, middle class Southwick zip up jumper car
draws up beside me
and glares at me with undisguised contempt.
He looks as though he would like to shout something
but no-one from Southwick talks to strangers
let alone shouts at them
so he just glares at me.

I stare back.
I don't glare.
I just stare.
I am very puzzled.

I check my person.
I am fully clothed.
My flies are done up.
My mud spattered T-shirt bears the logo
of an obscure folk band from Wigan.
Does everyone really hate the Tansads that much?

I am totally confused.
It's a beautiful Spring day
I am standing by the traffic lights
on a West Sussex A road
holding a bucket of frogspawn
and suddenly everybody hates me!
Another car hurtles past—
occupants screaming abuse.

Then the lights change again.
A car draws up beside me.
A very flashy, shiny one.
The boy racer inside is shaking his head.
He is gesturing to me
as though I am about to do something totally unacceptable
to something very important to him
and he really doesn't want me to.
I stand there.
I gaze at him in absolute bewilderment.

His window opens.
Then his mouth.
'Bloody squeegee merchant. Don't you touch my fucking car. Piss off and get a fucking job!'

I look at him in astonishment.
What has he just said?

Then I realise.

I am standing by the traffic lights.
I am holding a bucket.

He thinks I am about to start cleaning his car windscreen without his permission.

I walk over to the car.
I tip the bucket up slightly
and proffer the contents to him.
It is his turn to be confused.
He's a boy racer.

His car has a Romford dealership sticker on the window
He doesn't exactly have a herpetologist's soul.
He stares at the contents of my bucket.
He doesn't know what it is.
However, I think he realises that I'm not going to start cleaning his windscreen with it.

The lights change
he roars off
I walk back to the side of the road
and laugh till the tears run down my cheeks,
till my sides are killing me
till I pull a muscle.

* * * * *

It's March
eight years later.
My pond is full of spawn—
the great great grandtadpoles of that original bucketful.
As I crouch to look at the developing specks
the memory comes back.
I start to shake
the tears run down my cheeks
I nearly overbalance into the pond.
I run inside
I write this poem
I feel so happy.

Once Upon a Time

Gene Seibel

Once upon a time, mankind lived in caves and other stationary structures. Only snails and turtles carried their homes with them everywhere they went and all was well.

But man became restless and bored. He domesticated the four-legged animals and used them for travel. He could then set out on far journeys, but he missed his wife and his stone tablets. In order to kill two birds with one stone, he brought along his wife to carry his tablets and all seemed well.

Soon, woman became exhausted and angry. She tossed the stone tablets over a cliff and they rolled and rolled and rolled, and man said, "Behold the wheel!" Man tinkered, the four-legged animal was wedded to the wheel, and the chariot was born. Once again, all was well, but getting more complicated.

Man and woman roamed even farther, but they were cold and hungry and there were no motels or fast food joints. Woman said, "Let us stay home by the fire", and man once again became restless and bored. Man tinkered again, fire was coupled to the wheel, and the automobile came forth. Once more all was well—so long as there were tires, and gasoline, and insurance, and highways.

The automobile became a living being and multiplied, soon covering the face of the earth, and man and woman became trapped within the stomachs of the living beings. Woman said, "Let us stay home with our CD player and our telephone and our comfortable furniture." Man answered unto her, "I miss the comforts of home too, but we have car payments." Man tinkered yet again, and the automobile became married to the CD player, and the telephone, and all the comforts of home. Mankind thought all was well, but the automobile had become the SUV and the four-wheeled monster now ruled over mankind. And mankind said, "Oh to be like the turtle who is master of his house."

The Basho of Honk

Aaron Naparstek

There are only three
types of drivers—the morons,
the insane, and me

▷ ◁ ▷ ◁ ▷ ◁ ▷ ◁

Gruesome hit-and-run
fatalities up ahead
how awful—I'm late

▷ ◁ ▷ ◁ ▷ ◁ ▷ ◁

Our new minivan
so many cup holders it
needs a dishwasher

All poems are excerpted from Aaron Naparstek's book, "Honku: The Zen Antidote to Road Rage," Villard 2003. For more information and a place to write and contribute your own honku visit Aaron's web site: www.honku.org

Is it you or me
victim of insanity
honker or honkee?

▷ ◁ ▷ ◁ ▷ ◁ ▷ ◁

Natural rhythms
moons and tides supplanted by
yellow, red, and green

▷ ◁ ▷ ◁ ▷ ◁ ▷ ◁

Need a few days off
after Sunday-night drive from
the vacation house

▷ ◁ ▷ ◁ ▷ ◁ ▷ ◁

Just got my license
adulthood, power, freedom
let's go to the mall

▷ ◁ ▷ ◁ ▷ ◁ ▷ ◁

The evolution
of cars is stuck in reverse—
dinosaurs now rule

The air in Houston—
too darn thick to breathe but our
sunsets look pretty

▷ ◁ ▷ ◁ ▷ ◁ ▷ ◁

An SUV ad
featuring an F-16—
it's carmageddon

▷ ◁ ▷ ◁ ▷ ◁ ▷ ◁

Stop signs all four ways
since everyone is stopping
guess I don't have to

The Pedestrian

Ray Bradbury

To enter out into that silence that was the city at eight o'clock of a misty evening in November, to put your feet upon that buckling concrete walk, to step over grassy seams and make your way, hands in pockets, through the silences, that was what Mr. Leonard Mead most dearly loved to do. He would stand upon the corner of an intersection and peer down long moonlit avenues of sidewalk in four directions, deciding which way to go, but it really made no difference; he was alone in this world of A.D. 2053, or as good as alone, and with a final decision made, a path selected, he would stride off, sending patterns of frosty air before him like the smoke of a cigar.

Sometimes he would walk for hours and miles and return only at midnight to his house. And on his way he would see the cottages and homes with their dark windows, and it was not unequal to walking through a graveyard where only the faintest glimmers of firefly light appeared in flickers behind the windows. Sudden gray phantoms seemed to manifest upon inner room walls where a curtain was still undrawn against the night, or there were whisperings and murmurs where a window in a tomb-like building was still open.

Mr. Leonard Mead would pause, cock his head, listen, look, and march on, his feet making no noise on the lumpy walk. For long ago he had wisely changed to sneakers when strolling at night, because the dogs in intermittent squads would parallel his journey with barkings if he wore hard heels, and lights might click on and faces appear and an entire street be startled by the passing of a lone figure, himself, in the early November evening.

On this particular evening he began his journey in a westerly direction, toward the hidden sea. There was a good crystal frost in the air; it cut the nose and made the lungs blaze like a Christmas tree inside; you could feel the cold light going on and off, all the branches filled with invisible snow. He listened to the faint push of his soft shoes through autumn leaves with satisfaction, and whistled a cold quiet whistle between his teeth, occasionally picking up a leaf as he passed, examining its skeletal pattern in the infrequent lamplights as he went on, smelling its rusty smell.

"Hello, in there," he whispered to every house on every side as he moved. "What's up tonight on Channel 4, Channel 7, Channel 9? Where are the cowboys rushing, and do I see the United States Cavalry over the next hill to the rescue?"

The street was silent and long and empty, with only his shadow moving like the shadow of a hawk in midcountry. If he closed his eyes and stood very still, frozen, he could imagine himself upon the center of a plain, a wintry, windless Arizona desert with no house in a thousand miles, and only dry river beds, the streets, for company.

"What is it now?" he asked the houses, noticing his wrist watch. "Eight-thirty P.M.? Time for a dozen assorted murders? A quiz? A revue? A comedian falling off the stage?"

Was that a murmur of laughter from within a moon-white house? He hesitated, but went on when nothing more happened. He stumbled over a particularly uneven section of sidewalk. The cement was vanishing under flowers and grass. In ten years of walking by night or day, for thousands of miles, he had never met another person walking, not once in all that time.

He came to a cloverleaf intersection which stood silent where two main highways crossed the town. During the day it was a thunderous surge of cars, the gas stations open, a great insect rustling and a ceaseless jockeying for position as the scarab-beetles, a faint incense puttering from their exhausts, skimmed homeward to the far directions. But now these highways, too, were like streams in a dry season, all stone and bed and moon radiance.

He turned back on a side street, circling around toward his home. He was within a block of his destination when the lone car turned a corner quite suddenly and flashed a fierce white cone of light upon him. He stood entranced, not unlike a night moth, stunned by the illumination, and then drawn toward it.

A metallic voice called to him:

"Stand still. Stay where you are! Don't move!"

He halted.

"Put up your hands!"

"But—" he said.

"Your hands up! Or we'll shoot!"

The police, of course, but what a rare, incredible thing; in a city of three million, there was only *one* police car left, wasn't that correct? Ever since a year ago, 2052, the election year, the force had been cut down from three cars to one. Crime was ebbing; there was no need now for the police, save for this one lone car wandering and wandering the empty streets.

"Your name?" said the police car in a metallic whisper. He couldn't see the men in it for the bright light in his eyes.

"Leonard Mead," he said.

"Speak up!"

"Leonard Mead!"

"Business or profession?"

"I guess you'd call me a writer."

"No profession," said the police car, as if talking to itself. The light held him fixed, like a museum specimen, needle thrust through chest.

"You might say that, " said Mr. Mead. He hadn't written in years. Magazines and books didn't sell any more. Everything went on in the tomblike houses at night now, he thought, continuing his fancy. The tombs, ill-lit by television light, where the people sat like the dead, the gray or multicolored lights touching their faces, but never really touching *them.*

"No profession," said the phonograph voice, hissing. "What are you doing out?"

"Walking," said Leonard Mead.

"Walking!"

"Just walking," he said simply, but his face felt cold.

"Walking, just walking, walking?"

"Yes, sir."

"Walking where? For what?"

"Walking for air. Walking to see."

"Your address!"

"Eleven South Saint James Street."

"And there is air *in* your house, you have an air *conditioner*, Mr. Mead?"

"Yes."

"And you have a viewing screen in your house to see with?"

"No."

"No?" There was a crackling quiet that in itself was an accusation.

"Are you married, Mr. Mead?"

"No."

"Not married," said the police voice behind the fiery beam, The moon was high and clear among the stars and the houses were gray and silent.

"Nobody wanted me," said Leonard Mead with a smile.

"Don't speak unless you're spoken to!"

Leonard Mead waited in the cold night.

"Just *walking*, Mr. Mead?"

"Yes."

"But you haven't explained for what purpose."

"I explained; for air, and to see, and just to walk."

"Have you done this often?"

"Every night for years."

The police car sat in the center of the street with its radio throat faintly humming.

"Well, Mr. Mead," it said.

"Is that all?" he asked politely.

"Yes," said the voice. "Here." There was a sigh, a pop. The back door of the police car sprang wide. "Get in."

"Wait a minute, I haven't done anything!"

"Get in."

"I protest!"

"Mr. Mead."

He walked like a man suddenly drunk. As he passed the front window of the car he looked in. As he had expected, there was no one in the front seat, no one in the car at all.

"Get in."

He put his hand to the door and peered into the back seat, which was a little cell, a little black jail with bars. It smelled of riveted steel. It smelled of harsh antiseptic; it smelled too clean and hard and metallic. There was nothing soft there.

"Now if you had a wife to give you an alibi," said the iron voice. "But—"

"Where are you taking me?"

The car hesitated, or rather gave a faint whirring click, as if information, somewhere, was dropping card by punch-slotted card under electric eyes. "To the Psychiatric Center for Research on Regressive Tendencies."

He got in. The door shut with a soft thud. The police car rolled through the night avenues, flashing its dim lights ahead.

They passed one house on one street a moment later, one house in an entire city of houses that were dark, but this one particular house had all of its electric lights brightly lit, every window a loud yellow illumination, square and warm in the cool darkness.

"That's *my* house," said Leonard Mead.

No one answered him.

The car moved down the empty river-bed streets and off away, leaving the empty streets with the empty side-walks, and no sound and no motion all the rest of the chill November night.

Sherlock Holmes and the Case of the Stolen Bicycle

Scott Munn

After my bicycle was stolen on Tottenham Court Road during a real London particular, I made the long walk to Baker Street, for despite the unimaginativeness of the crime, I had hopes that the Great Detective would take my case if only because the chances of retrieval of my beloved mount were so very slim and I've often read that he likes a challenge.

Mrs Hudson answered the bell and ushered me up into the sancto sanctum sanctorum—not before I'd noticed the curtains fluttering up above. Once inside, Watson immediately put me at my ease with a jolly story about the Afghan campaign, quieting only when Holmes developed a tic and muttered something about "that damned Jezial bullet" in an uxorious tone. Truly they were like an old married couple. In the silence that followed I felt Holmes' scrutiny most keenly. His eyes were insatiable, feeding gigabytes of information to that vast computer of a brain as fast as they could alight on each minute detail of my countenance. I must admit I squirmed a little, nervous that the intrusion might penetrate my very psyche as if I were sitting on the couch of Mr Sigmund Freud, and leave me feeling picked clean like the bones of a fish. Even Watson grew restless after a spell.

Suddenly Holmes unfolded from his crouch and sprang out of his chair, almost knocking over his pipe. "Come, Watson, the game is afoot," he ejaculated, and the two of them swept out of the rooms. I found this most singular and not a little rude. The landlady kindly ushered me to the door, tutting all the while.

Several weeks passed without word from Baker Street. Even my emails went unanswered, save for a polite autoresponse. Or perhaps it was just my impatience which prompted such unkind thoughts. Oh, I admit it. I was an emotional wreck. I'd heard stories about what can happen to stolen bicycles that would make a grown man weep. It seemed that I'd caught my own Jezial bullet.

Presently there came a knock on my door. I extracted myself from the tangled detritus of the nest I'd made—mostly bicycle bits and bobs which brought me comfort, but also a fair amount of oriental takeaway containers—and went to greet my unwelcome caller.

He was a most peculiar-looking man, humpbacked yet militarily straight, with a single arm several times too long for the rest of his eggcup-shaped body, the complexion of a charred beetroot, nose like a weathercock and the scalp of a sheared rabbit. He was also quite maladroit in the manners department. He scurried in before I could bar the way and ululated most remarkably for a minute before settling down on my favourite settee.

"What on earth is the meaning of this intrusion?—" is as far as I got before my visitor broke into the outraged peroration I had planned with a cackle and threw off his disguise with a flourish. It was Holmes himself! I fairly collapsed into my nest, tapped out in every department, whereupon Watson swept into the room, immediately assessed the situation like the medical professional he is, gave Holmes a severe look (which the latter studiously ignored), and put me to bed with a cup of broth.

When I awoke they were both gone, but a loitering Baker Street Irregular was waiting at my door to hail me a cab. Presently I found myself once again in the company of Holmes and Watson. "Do sit down, my good fellow," said the doctor. As I was taking his advice Holmes gestured to the Irregular, still awaiting his silver, whereupon the little scalawag absconded from our presence momentarily and just as suddenly reappeared, carefully steering my bicycle into the room. Needless to say I promptly fainted once again.

"—inexcusably dramatic" I heard Watson scolding Holmes as I emerged from the mental fog which had minutes before enveloped me. I looked at them both. "Explain," I croaked, whereupon Holmes fairly rubbed his hands together with glee.

"It was perfectly obvious to anyone with eyes to see and an analytical mind that you had recently suffered a loss of a most devastating kind," he began in preamble, dispensing with his pipe and declining the cocaine offered by his Irregular supplier, whom he now shooed away. "I've seen that look a thousand times and recognized it as such even as you stood, a broken man, outside this window a few short weeks ago.

"Even if you hadn't been wearing your cycling mitts" (they were on my hands still) "there was every indication that you are of the velo persuasion. I won't bore you with all the details, but will leave it to Watson should he ever put pen to paper in another of his fanciful tales"—this followed by a hurrumph from the wounded party.

"Suffice it to say the fact that you stopped and checked the way for other cyclists before crossing Baker Street, rather than blindly blundering into that mighty avenue like most pedestrians and motorists temporarily unawheel, was a thunderingly obvious clue.

"Bare seconds after Mrs Hudson escorted you to my study I had deduced that your missing bicycle was of the hybrid variety, charcoal in colour, with foam grips, bar-ends, a lowrider rack, Lumocycle Standlicht dynamo lamp, equipped with toe-clips (uncinched). These observations were really all quite elementary. As you know, when you have eliminated the impossible whatever remains, however improbable, must be the truth. Moriarty is behind the theft. He sits as does an evil spider in his web. I fear he has stripped your poor Marin's vital components for resale before offloading the empty husk to an unsuspecting party in another town. It was easy to conclude that even I would not be able to retrieve your stolen bicycle in a state in which you would wish to find it—although I did hear talk of it in Lewes, where the disguise you saw allowed me access into that dark world—and it would be far easier and kinder to simply buy you a new one. After perusing the agony columns I came across an ad from our local bike shop and sent Watson to fetch a similar beast. It appears the doctor has erred on the side of generosity in his task."

It was of course true. In my excitement as the Irregular had wheeled the bike in I had singularly failed to take note that it was nothing like that which had been stolen. It was even a different colour. I could not thank them enough. As I collected my new steed and prepared to leave Baker Street, it occurred to me that the great detective might be able to help me with another matter:

"A friend of mine possesses a bicycle which evinces a peculiar 'ticking' sound emanating from the vicinity of the bottom bracket—"

As Holmes' face assumed a rictus of undiluted horror, Dr Watson bundled me out of the study and onto the street. "Terribly sorry old chap," he said. "There are some cases that even Sherlock won't touch."

Bending a Straight Road

Peter Gelman

With an illustration by Neal Skorpen

As I rounded the bend, huffing along New Lower River Road, I suddenly stopped whistling. The reason was—well, there was a subtle reason that you may not understand, and I am sort of busy. But regardless of the reason, my whistle cessation was in itself noteworthy. Not only was the whistle cessation noteworthy, but so was the actually whistling that preceded it.

I had begun that round of whistling in part because of a nervous vibration. Vibrated and agitated for absolutely no reason, I wanted to whistle my tune not twice, not zero times, but one pure, true, naïve little time. The problem was that I couldn't be sure if I had already whistled that vitally imperative tune when I biked across the Columbia River bridge. The thunderous traffic (that was but a symptom of that cosmically nervous vibration) obliterated even the memory of my lips forming an "O". It did so with numerous "broom", "mrah", and "nyah" sounds, which as you know, are not conducive to pure, sweet bouts of happy whistling.

Here, however, the road was quiet, and flat. I pedaled

happily with my special cha-cha motion. The heavy metal tune ("—- —- – —-") validated me. It showed me that I was still—even now—still in touch with the Great Unwashed. I pedaled, lips forming the obligatory "O". My nostrils, too, formed two soft French vowels with their accent marks; you needn't know more than that. I felt myself comforted by a sort of shrink wrap clinging to my soul, a protective shrink wrap of popular culture.

And now that we touch upon the subject of my soul, viewing it in perversely erotic manner through its tight covering of cellophane, we approach the point where I stopped whistling and the outrageous "incident" began. Very well. I will blushingly tell you that I stopped whistling because upon making a private inventory I noticed something else—there was a tear in that shrink wrap. And shrink wrap with a tear is as good as no shrink wrap at all. My lips stopped forming an "O"; they now formed a sort of dash (or perhaps the "equals" sign), and I listened.

I heard my tires. Like a sensitive college woman with a guitar on Open Mike Night, my tires were quietly singing a new song of their own invention. You know how easy it is to fall in love with a young woman singing about big, hopeful things and expecting to change the world, a world known as Planet Earth. Granted it is not as easy to fall in love with a bicycle tire, however so fresh and knobby. Naturally I forgot all about my torn shrink wrap, and that may explain some of the unfortunate events that followed.

It was sunny, two days after Thanksgiving. Inspired by the digestive euphoria of three days of "tofurkey" sandwiches, I put new 100-PSI tires on my bicycle. My tires hummed happily and I listened, holding my head sideways.

In thus a posture I looked left at the flat pastureland with a bull watching me back. I saw crows in cottonwood trees and a hawk circling with sparrows chasing it. Then I looked right at the lake through the trees. I saw a heron land with a clumsy splash that annoyed the ducks.

With such things occupying my attention, I didn't see any hints or warning signs that the sky was cracking open. But when I turned my gaze from the quacksters to the sky

I did see something that told me that the great blue dome had cracked open. Something had broken the laws of Nature: A huge white triangle hung in the air.

I blinked, and it went away, until I opened my eyes again. There it was, hovering in all its impudent, stubby, obnoxious, looming triangularosity. It seemed enormous, miles and miles of floating, three-pointed sky-cracker. It was beyond mortal mathematics to fathom its hypotenuse.

It made no sense; I felt all wobbly and stopped the bike without my habitual flourish of the hips and elbows. I stared at the impossible thing. Although I suddenly knew it must be Mount Saint Helens, the momentum of my confusion caused me to wonder, does my dependence on the sky not cracking open make me unnecessarily vulnerable to cosmic changes? After all, who knows? The sky might crack open someday. Personally, I would prefer the Great Blue to wear a bike helmet and not crack open.

A bank of clouds tinted the same color as the sky obscured the midsection of the mountain. It was a trick of horizon atmospherics, the curving horizon playing with light and mass. I took a moment to smile and marvel. If we bicycle on Earth, we bicycle on an enormous sphere. Lines on a sphere (illustration: plum), such as roads, aren't straight like a carrot, celery stick, or Boy Scout. They're curved like a banana. But we think of them as straight in our two dimensional folly.

That is what I thought, as I resumed biking that Washington road: "Isn't it strange that we cyclists roll and roll, pushing the Earth backward under our tire, and feel that it is we that move forward, and not the planet moving backward?"

"Turn, planet, turn!" I cackled. "I'm turning you backwards you little hub bearing!" I hopped my bike over a crack in the road, delighting in the doughtiness of my Antelope.

As I began to pedal with just my right foot, scratching my left against the handlebars, I considered, "Every road is on a sphere, namely, a planet—a planet called Earth. Every road, no matter how straight-looking, is secretly curved. All I have to do is find the way it's curved. So that must mean every road, even up a mountain, must be, in a certain sense, downhill. All I need to do is find that secret downhill quality to even the uphill roads and I will win every race on this planet, a planet known as Earth."

And then I considered the selfishness of this thought. After all, if I was busy winning every race, spraying champagne on my head, posing for photographs, kissing beauty queens and fondling sponsors, how would I possibly save the Earth at the same time? It seemed reasonable that I could only do one or the other—*not* both.

With my ambition properly chastened, I rolled by the sign that marked the road's entrance into the nature preserve. You could tell it was a nature preserve sign because it was full of bullet holes. It made me a little sad that I too, did not have a musket (or better yet, a blunderbuss) with which to express my love of the 2nd Amendment. Alas, my hands have too long habituated themselves to the handlebar curl, which leaves me good for little else, except perhaps air guitar.

Cycling on, my tires bumped over the remains of last July's roman candles and bottle rockets. Suddenly I slowed and stopped to say, "Hi there, little snake-person."

Sunning its back on the warm asphalt was a striped garter snake about seven inches long.

Then I asked it, "Pretty little snake, want me to put you in the cold grass or you want to continue sitting where you are getting your belly nice and hot?"

The reptile's tiny forked tongue darted in and out, which made me think I was bothering it, and therefore I should leave it alone. If it were bothering me, I would, of course, have had the right to bother it back; that is the basis of all moral law.

My thoughts continued: "Should I move it off the road? Yeah. Nah. Maybe—yeah. Nah. I don't know."

In replacement of making a decision, I crossed two ancient sparklers into an X. The X would mark the spot where I had befriended the snake, so I could look for it on my way back. I biked on happily.

Behind me I heard a horrible noise, but I wasn't even

curious. That was because this hellish, horrible noise was the ubiquitous and everyday sound of an overpowered automobile. It hurt my ears and tore my concentration but I didn't even flinch. Eventually I allowed myself to glance, bored, at the boring pickup truck pulling a tedious trailer with four dull ATVs. The truck roared, pouring black exhaust into the air, and I rolled my eyes at its banality. The trailer behind it rattled and squeaked. I yawned to show my disdain.

I didn't ask it to, but as I yawned, my brain churned out a new thought. Annoyed at my brain, I stopped pedaling, and dragged my toes. Eventually my momentum stopped, and I turned the bike around.

Finally, I came back to the sparklers. The sparklers, I noticed, remained in their perfect X. Nearby I found the snake, its middle section curling and uncurling a little.

After watching the snake's slow movements I noticed something and looked away. I put my arms into a restful Tai Chi posture. I tried to make myself think that the snake was still alive and I was mistaken about it having its head run over.

Suddenly furious, I grabbed the edge of the asphalt with my fingers, and tried to whip it like a rug so that the truck, far ahead now, would go flying into orbit and be run over by Mir. It was a silly thing to do and of course I could not do it.

When I straightened up, however, I noticed a change. Where the road had been straight before, it now curved. I pursed my lips, shrugged, and cycled back toward Vancouver.

I sensed something—I'm not sure what it was I sensed, and I'm not saying that what I sensed was a something. Perhaps it was just a stirring in the general indifference of the global atmosphere.

The stirring was actually a rotation. The rotation was in fact a man's feet going around and around. I blinked and looked again. I was astonished to see in my rear view mirror a man floating in the air. He was a cyclist, tucked down and pedaling hard. But he had no bicycle.

The stranger caught up with me and paused beside me for a moment. Now I could see the bike's strange, bridge-like framing, with diagonal crosspieces holding the rider in a sling-seat.

It was a Pedersen bicycle! Invented by a bridge engineer in 1893, the design applies bridge strength to something with lightness and mobility. The machine is still a unique intimation of the 19th century promise of Progress.

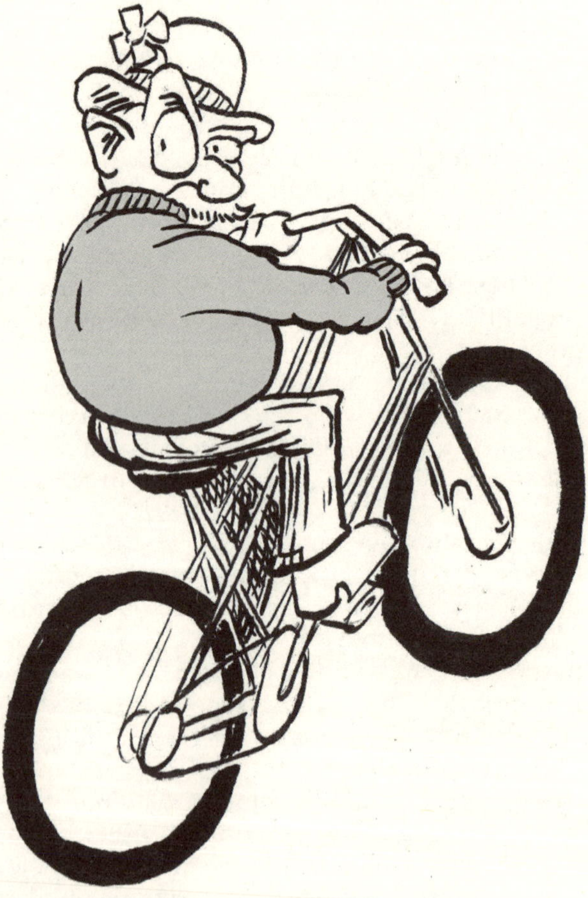

"Good-day," the cyclist offered. He was formally dressed in a gray woolen turtleneck and matching knickers. He wasn't even breathing hard through that enormous walrus mustache. "Care for a race?"

There wasn't anything unpleasant about him at all—quite the contrary. And yet, I didn't like him, because I was afraid of him.

Feeling that thought as I rode alongside him, I smiled to myself—how silly! There was nothing wrong with him at all; he was friendly. So what if he rode an admirable antique bicycle? He kept his mustache well combed and free of crumbs. Did it bother me that he pedaled at a robust rate without breathing hard? It didn't bother me, and why should it?

It's true that he might have seen me let the truck run over the snake. So what if he came to chase me down, to make me crash, to drain my blood and pierce my eyes with sharpened bicycle spokes? Who was I to let my small objections lead me to say no to his challenge? I had new tires and I was not going to let even a demon think he could cycle faster than me.

So I replied, "Sure. I'll say 'ready, set, go.'"

I was smart enough to know that I did not want to have my eyes poked with sharpened bicycle spokes. I was so clever as to think that if I could beat him, the snake's death wouldn't be my fault. Thus I took the sensible precaution of accelerating ten yards ahead of him and up-shifting to my big ring *before* I blurted out, "ready-set-go!"

As he realized that I had cheated him out of a fair start, the stranger's smile disappeared and his mustaches shivered with anger. The stranger flung after me effortlessly.

Even with the rider's upright posture, the strange bike passed me, gently clicking, quite close by. I watched in awe —and then in fright as his leg brushed my leg. As I fumbled, he passed me. The distance widened. It was clear I could not keep up with him.

I felt no amusement. I shared no bicycle fraternity. I felt cruel, mean, and streamlined. The wind made my eyes tear. I felt awful, spiteful, and full of spit which I quickly spat. My mouth was dry, why did I spit? I felt greedy, angry, and lumpy on the seat, so I stood up to sprint. I felt malicious, grotesque, and spastic, and almost fell over, so I sat down again. It was nice to sit on a saddle and coast. Comfort embraced me.

The Pedersen cyclist zipped farther ahead. My mood changed, now. Now I felt ticklish, handsome, and smug, so I said, "Whatever."

That wasn't sufficient, so I tried again to achieve the professional disdain of an adolescent girl: "*What*-ever." And yet, voicing disdain did not dispel the thought that I should have gently urged the little garter snake out of the warm road. Was it evil of me?

No! Evil, as I defined it, was nothing less and nothing more than someone cycling faster than me. I could not permit it!

And yet, I felt tired, numb, and studious, so I studied my opponent. He was my rival. My villain. My dastardly race opponent. He was nothing less than a sling-seated evildoer in a turtleneck sweater!

I could not pedal any faster, so there was only one thing to do—and this idea, you will agree, was absolutely brilliant. You can believe me because I have so many brilliant ideas that really they just seem like average ones to me. I wish I had a special flyswatter that could swat out just a few of my brilliant ideas; it's such a bother having so many all of the time. So you see, I was actually glad to use one of my ideas in this manner, because it relieved a bit of the dazzle-pressure inside my head.

My idea was simple: all I had to do was *slow down*, and *bike in another direction.*

You see, I didn't need to chase *him*, because I wasn't trying to catch him to murder him. Did he want to poke out my eyes with sharpened bicycle spokes? Well then, he'd have to catch me first. However, at this moment, I was cranking hard in pursuit of him. I was chasing him. Why not turn around, and make him chase me?

It was true that he seemed to be a much faster cyclist than me. Nonetheless, if he chased me, he would have to let me go first—that was the definition of chasing.

Furthermore, letting me go first, he would have to let me bike faster than him. I would not only escape Hokey Pokey Bike-Spoke Pokey, I would win the race.

Any way he handled it, I had out-smarted him.

Braking hard, I caught him unawares, and he shot ahead.

Skidding, I yelled "Ha!"

Immediately, he, too, braked—and the Pedersen just stopped moving with devilish abruptness.

I, however, skidded with a sound like someone breaking through the wicker of an antique chair in a store with a sign saying "Please don't sit in the chair." My rear tire lightened and lifted, pressing me upward. I pushed myself off the saddle, moving my weight backward. The rear wheel lifted more, and I felt my balance shifting—I was about to fall forward, over the handlebars, and fling my face into the spokes of his rear wheel.

The thought occurred to me: "I am having a thought!"

(And then the actual thought followed—)

"I need to move my center of mass backward, and fast."

I quickly concentrated all of my deep, heavy philosophical thoughts about saving the world at the bottom of my brain stem. This displaced my lighter, more buoyant and frivolous thoughts such as how to button my shirt buttons in the right button-hole. Still I tipped! So I added my lightest of all effervescent thoughts, reliable ways to earn a living, which cheerfully rose up like cheap champagne bubbles into my throbbing forehead.

Thus lightened in the front, I did not fall forward. My momentum stopped, and my rear wheel came down.

Even as this happened, I relished the thought of my rival smarting under my outrageous "Ha!"

Slowing, I skidded alongside the Pedersen cyclist. As my momentum stopped, I suddenly changed my mind. I realized I was tired, that I did not want him to have an easy time drafting me, that I would rather draft him. I quickly turned my front wheel and balanced, with no forward motion.

He did the same. We stood in our track stands, quivering with the effort of maintaining motionless balance.

"Did you 'Ha!' me, sir?" he asked firmly as he balanced there beside me.

"What if I did?"

"I ask you again, did you 'Ha!' me, or did you not?"

"I've been known to 'Ha!' spontaneously from time to time," I explained, "and if anyone overhears it, well then, they can take it as they please."

He considered this; I felt my rhetorical arts had quite demolished him.

Then he turned around in his sling-saddle, quivering as he balanced on the motionless Pedersen. Staring at me, he suddenly cried out: "Ho!"

Then he began to move again. I pursued him through the rushing air as the infamy of his "Ho!" echoed around and around in my brain. As we turned onto Hillhurst Road, headed north to Ridgefield, I tightened my hands on the grips.

His "Ho!" made me quite angry indeed. Therefore I told my leg muscles I would accept no pleas of mercy, no nerve impulses of pain indicating a build up of lactic acid or of any other kind of acid, or any other kind of lactics. My legs obeyed and my feet turned two tornadoes around the crank. The air tore at me, clutched my face, made my earlobes flap.

My antelope machine accelerated so that I was able to tuck my front wheel in the immediate space behind his, where the air suddenly became quiet.

We zipped over North Fork Avenue, passing farms, cows, shooting ranges, and egret nests. We zapped through the town of La Center. I grinned, confident of victory: I was drafting him. It took me no effort to keep up; soon I would attack.

As if sensing this, he picked up even more speed. Faster and faster he flew. The aerodynamic pocket sucked me along and whipped me faster than I could ever pedal myself. Miles began to flutter by, alternating with the occasional kilometer.

When I saw how close we had come to Rainier, I became concerned how far north I was going. I would have a hard time finding my way back to Portland. After some hesitation, I decided to let him win.

I squeezed the brake levers and my wheel rims *screamed.* But I did not slow down. I braked harder, and fire flew from the screaming brake pads, but my velocity did not decrease. Not liking that sound and combustion, I eased off.

Now I just tried to hang on. I don't know how fast he was going but had already come to the outskirts of Tacoma. We blasted through the Tacoma streets beside cars of all nations. We serpentined along the narrow road with occasional views of SeaTac between the trees. Then we rocketed up a long, sloping hill, and I saw the shipyards and the distant bridges of a gigantic automobile-choked city. We rolled even faster down the hill toward its center.

I realized the only way I would regain my freedom was to beat the Pedersen rider. I groaned as I somehow pedaled harder. Just as we passed the Boeing complex I began to catch up to the stranger. As we entered downtown Seattle I caught a glimpse of him looking back at me, wiggling his eyebrows in victory. I remembered the way he called out "Ho!" so insultingly and decided to refuse him this pleasure. I grit my teeth and charged until I was almost even with him. He opened his mouth with surprise, and then shut it angrily.

I discovered he was now trying to physically block my way ahead. He ducked through narrow gaps in the angry crowds of people—people holding signs—shouting and singing—someone wearing a turtle costume—the Pedersen weaved through the befuddled men in suits—clutching their briefcases and holding handkerchiefs to their mouths —and toward the row of black-armored policeman and, faster and faster, with vertiginous speed, toward one officer with intense green eyes holding a "beanbag" gun, the barrel growing larger and larger I followed the Pedersen rider down into the darkness of the gun barrel which exploded with nauseous, beanie energy.

I woke up later, and realized that I had crashed. I lay tangled with the frame of my bike.

A stranger's hands grabbed me! I grimaced and turned my head from the sharpened spokes!

But no, it was just the gentle hands of a tearful, red-faced woman with multicolored yarn braiding her hair. I wondered if there had been a sweater-knitting incident. Coughing, wheezing, and defiantly chewing an organic macaroon, she helped me stand up. It was hard to breathe—something in the air tasted like fire and poked my eyes. As she pulled me up, a small lead-filled bag slid off of my chest. Before I could thank her, she left, leaving a tantalizing scent of coconut and almonds.

Flizzzz! A Pedersen whizzed by on my left! I looked to my left, but instead of a man on a Pedersen bicycle I saw a police officer on horseback galloping behind some people who were running away and screaming. I assumed, of course, that the police officer was concerned and trying to help them. It impressed me how he, like the Queen of England, insisted on bestowing on people the honor of knighthood. To do so, he applied his ceremonial staff (or billy-club, to use the vernacular) on either side of the knighted person's head.

Zlarroo! A Pedersen scorched by on my right! I looked to my right, but instead of a Pedersen I saw a carpenter was drilling screws in plywood to clasp it over a boutique's window. This, I guessed, reflected the fact that internet shopping was replacing actual physical boutique shopping. That furthermore explained why the armored police officers were "pointing and clicking" their cans of mace at the shoppers on the street.

Believe it or not, I wasn't yet done looking. I looked straight ahead and saw, through a whorl of strange white smoke, a straight road.

I straightened the handlebars of my bike and mounted it. I began the long ride south to Oregon, and vowed never to bend a straight road again.

Sadly, I didn't keep that vow. I bent several small, inconsequential roads on my way back.

The Persistence of Nature and the Nature of Persistence

Scott Massey

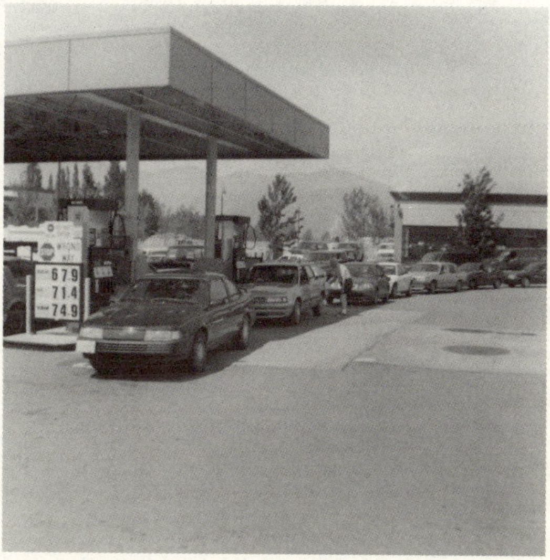

"Tunnel vision is a disease in which perception is restricted by ignorance and distorted by vested interest."
— Tom Robbins

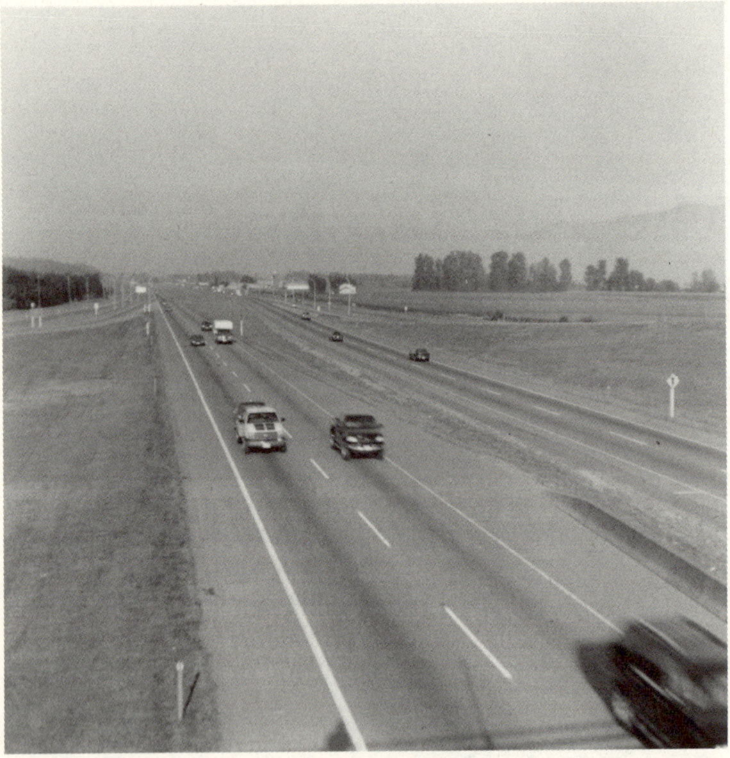

Never before has there been a species so conscious of its own existence and yet so oblivious to its own demise — Scott Massey

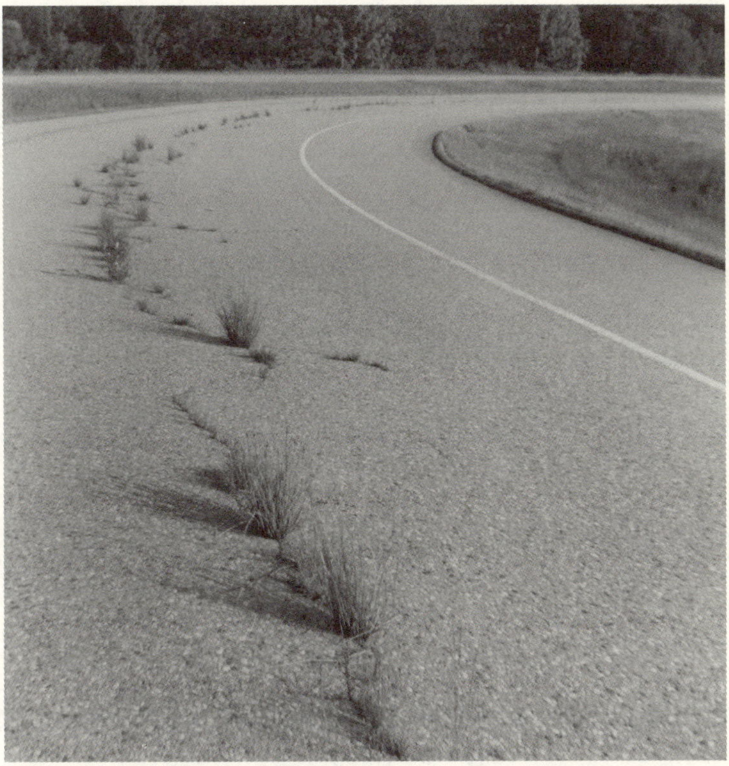

"Grandeur is seen to fade, but grass springs eternally." —
John Clare

Salvation through Car-toons

Andy Singer

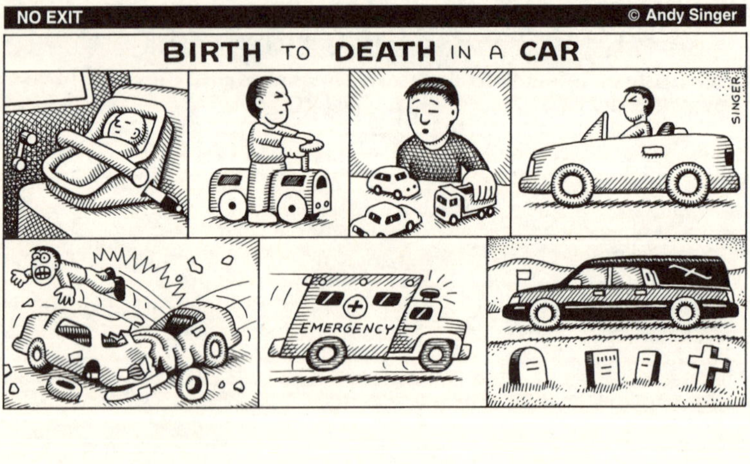

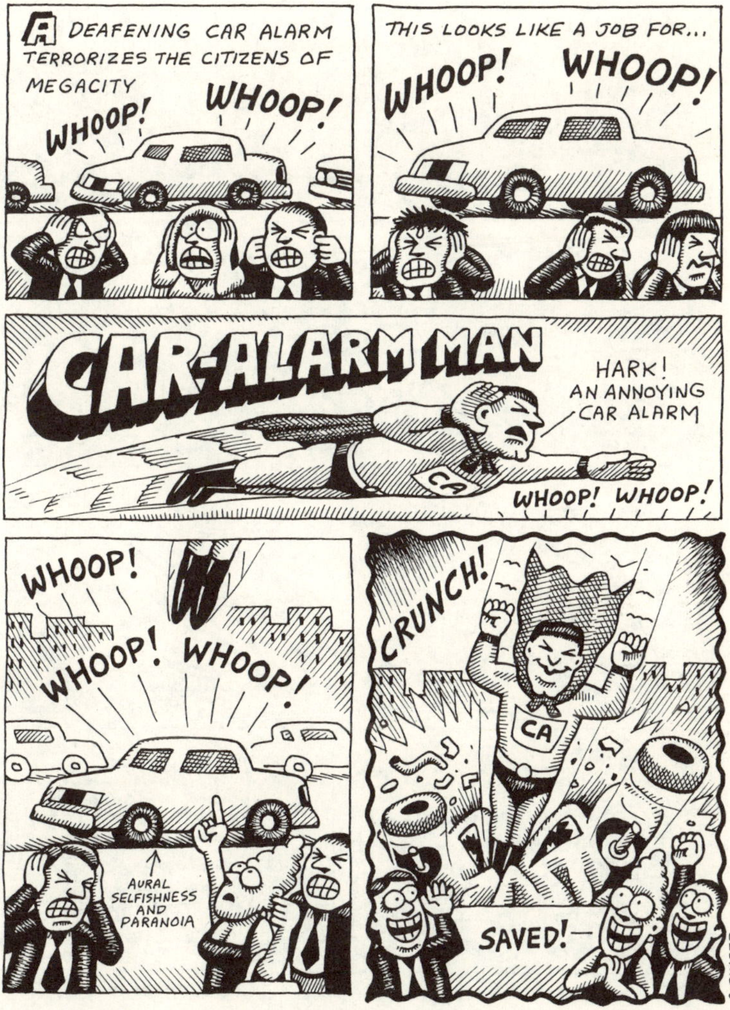
A DEAFENING CAR ALARM TERRORIZES THE CITIZENS OF MEGACITY
WHOOP!
WHOOP!
THIS LOOKS LIKE A JOB FOR...
WHOOP!
WHOOP!
CAR-ALARM MAN
HARK! AN ANNOYING CAR ALARM
CA
WHOOP! WHOOP!
WHOOP!
WHOOP!
WHOOP!
AURAL SELFISHNESS AND PARANOIA
CRUNCH!
CA
SAVED!
A. SINGER

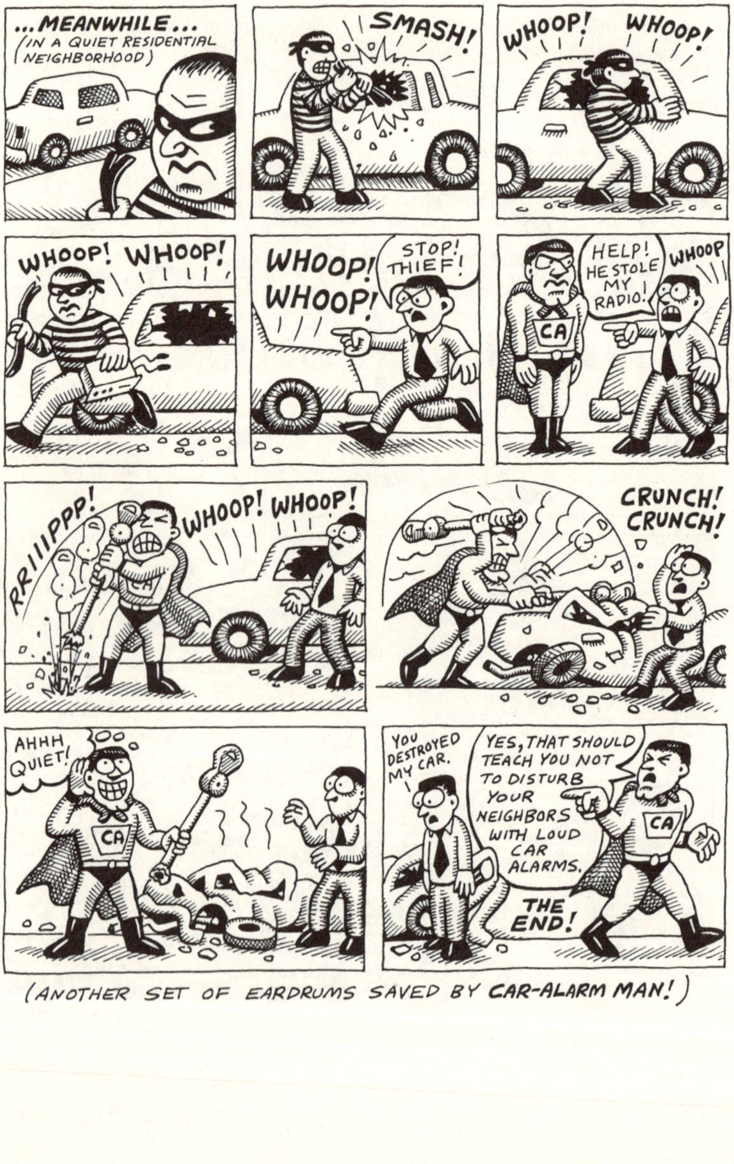

(ANOTHER SET OF EARDRUMS SAVED BY **CAR-ALARM MAN!**)

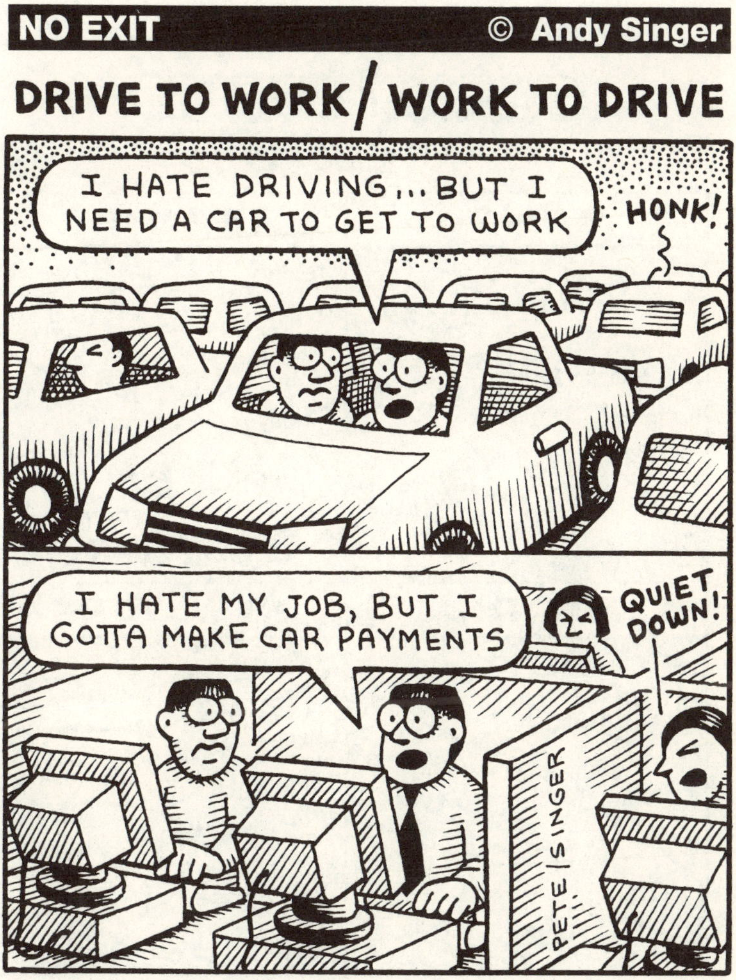
NO EXIT
© Andy Singer
DRIVE TO WORK/WORK TO DRIVE
I HATE DRIVING... BUT I NEED A CAR TO GET TO WORK
HONK!
I HATE MY JOB, BUT I GOTTA MAKE CAR PAYMENTS
QUIET DOWN!
PETE SINGER

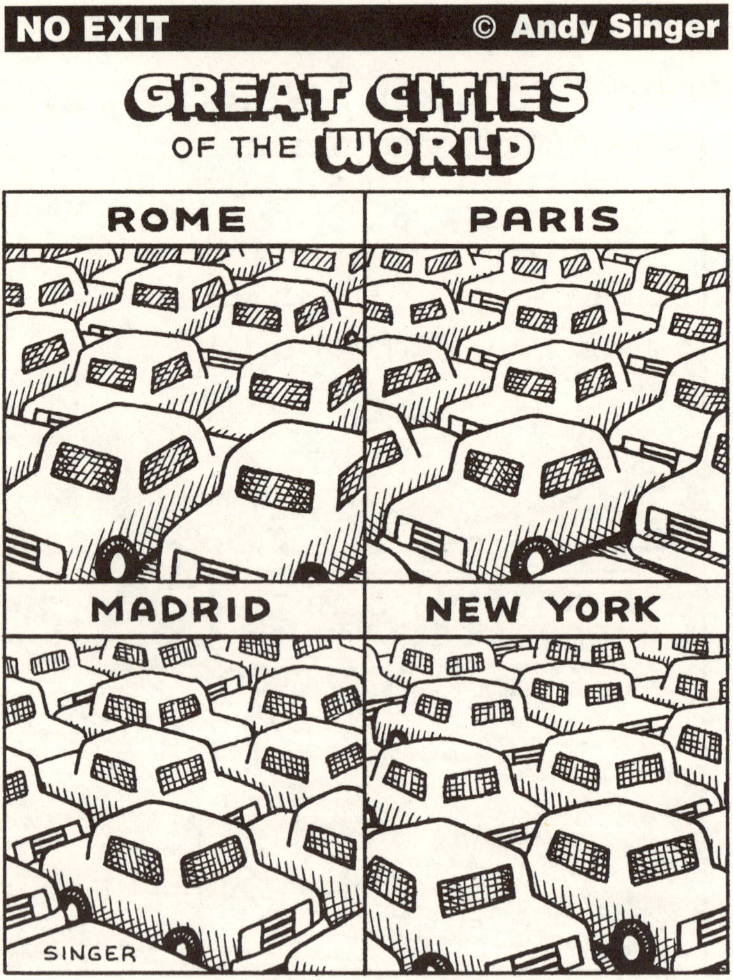
NO EXIT
© Andy Singer
GREAT CITIES
OF THE WORLD
ROME
PARIS
MADRID
NEW YORK
SINGER

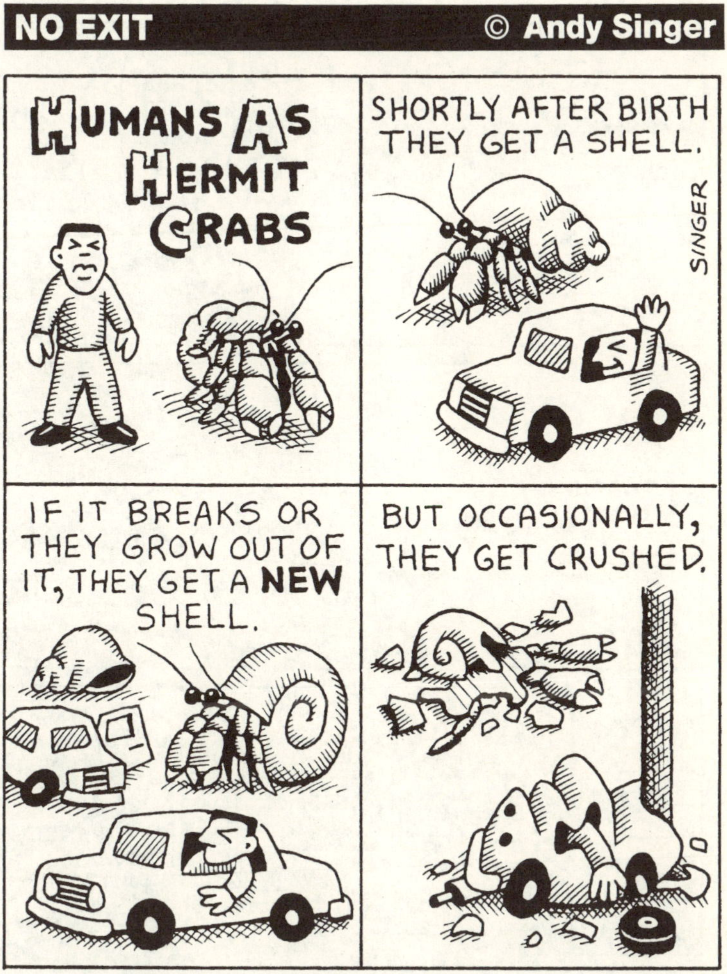
NO EXIT
© Andy Singer
HUMANS AS HERMIT CRABS
SHORTLY AFTER BIRTH THEY GET A SHELL.
SINGER
IF IT BREAKS OR THEY GROW OUT OF IT, THEY GET A NEW SHELL.
BUT OCCASIONALLY, THEY GET CRUSHED.

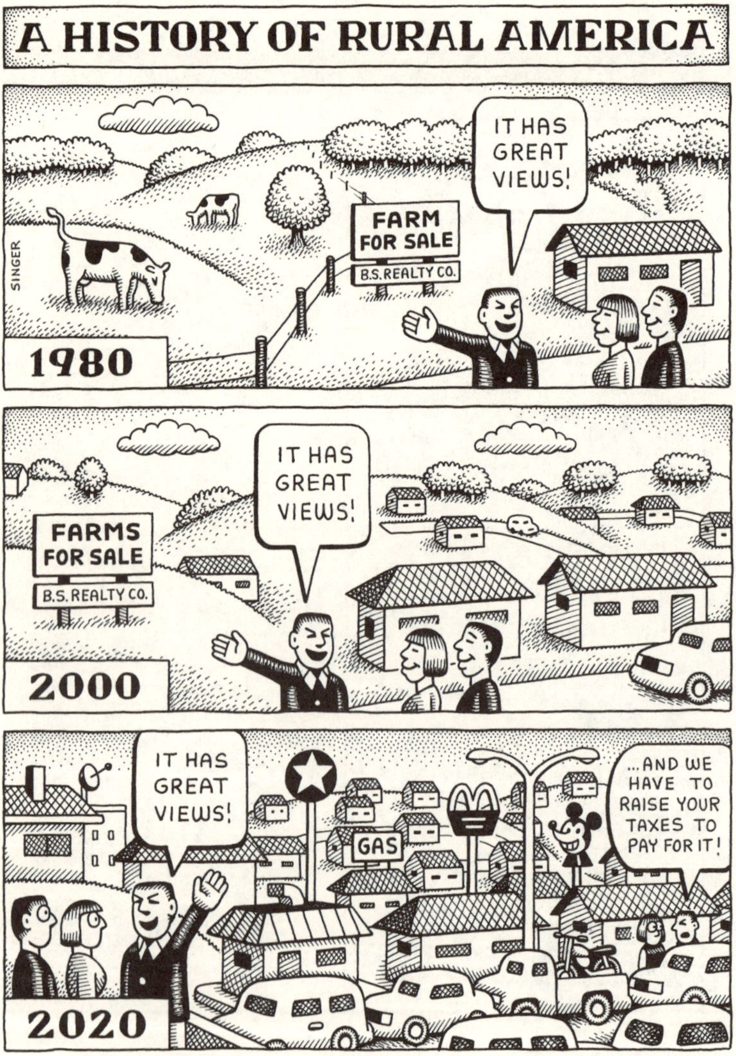
A HISTORY OF RURAL AMERICA
SINGER
FARM FOR SALE
B.S. REALTY CO.
IT HAS GREAT VIEWS!
1980
FARMS FOR SALE
B.S. REALTY CO.
IT HAS GREAT VIEWS!
2000
IT HAS GREAT VIEWS!
GAS
...AND WE HAVE TO RAISE YOUR TAXES TO PAY FOR IT!
2020

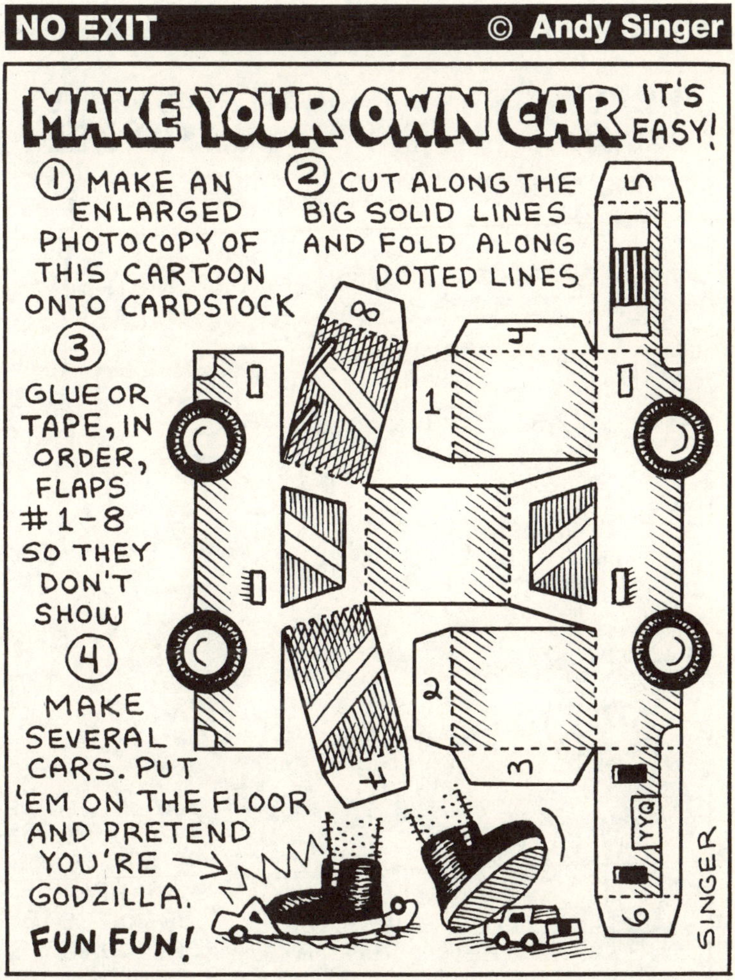
NO EXIT
© Andy Singer
MAKE YOUR OWN CAR IT'S EASY!
① MAKE AN ENLARGED PHOTOCOPY OF THIS CARTOON ONTO CARDSTOCK
② CUT ALONG THE BIG SOLID LINES AND FOLD ALONG DOTTED LINES
③ GLUE OR TAPE, IN ORDER, FLAPS #1-8 SO THEY DON'T SHOW
④ MAKE SEVERAL CARS. PUT 'EM ON THE FLOOR AND PRETEND YOU'RE GODZILLA.
FUN FUN!
1
2
3
4
5
6
7
8
YYQ
SINGER

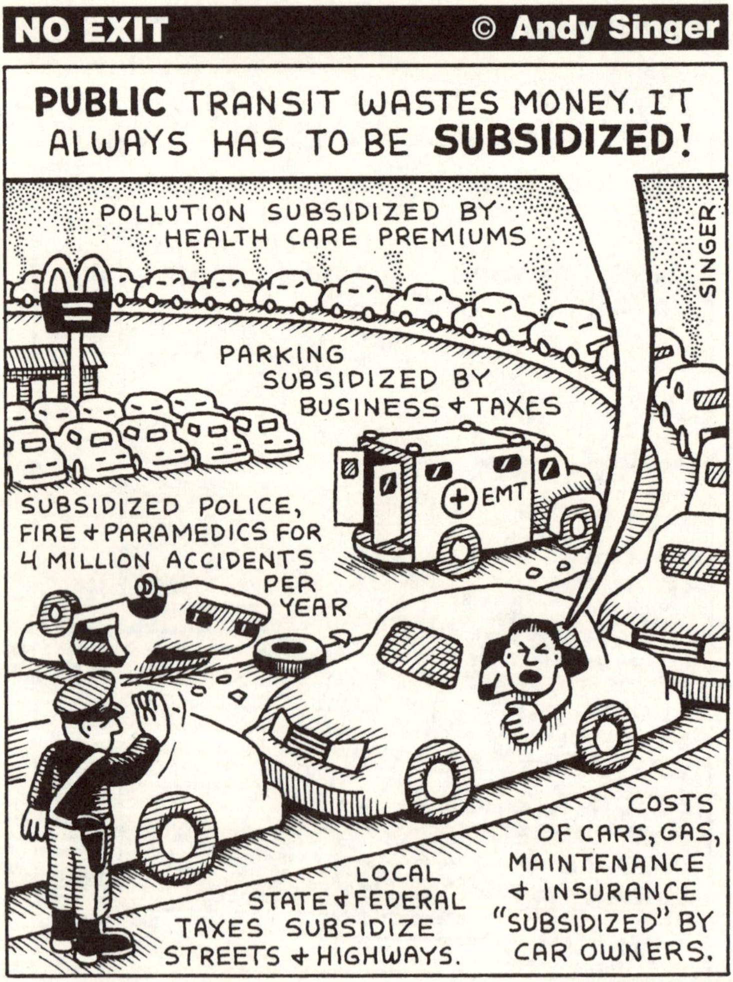
NO EXIT
© Andy Singer
PUBLIC TRANSIT WASTES MONEY. IT ALWAYS HAS TO BE SUBSIDIZED!
POLLUTION SUBSIDIZED BY HEALTH CARE PREMIUMS
SINGER
PARKING SUBSIDIZED BY BUSINESS + TAXES
EMT
SUBSIDIZED POLICE, FIRE + PARAMEDICS FOR 4 MILLION ACCIDENTS PER YEAR
LOCAL STATE + FEDERAL TAXES SUBSIDIZE STREETS + HIGHWAYS.
COSTS OF CARS, GAS, MAINTENANCE + INSURANCE "SUBSIDIZED" BY CAR OWNERS.

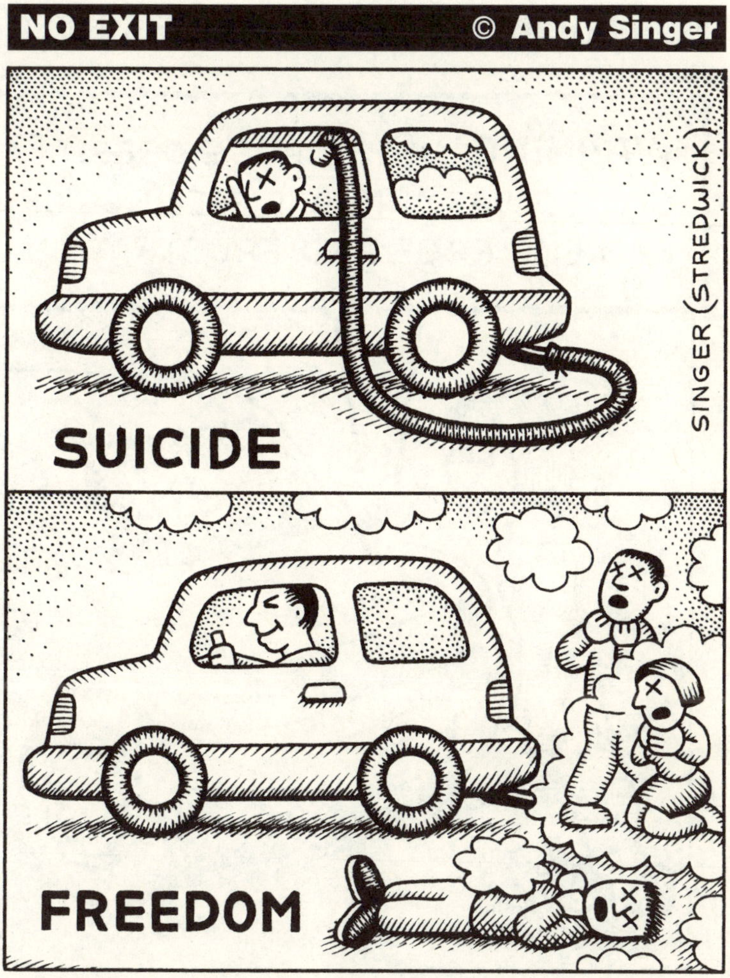
NO EXIT
© Andy Singer
SINGER (STREDWICK)
SUICIDE
FREEDOM

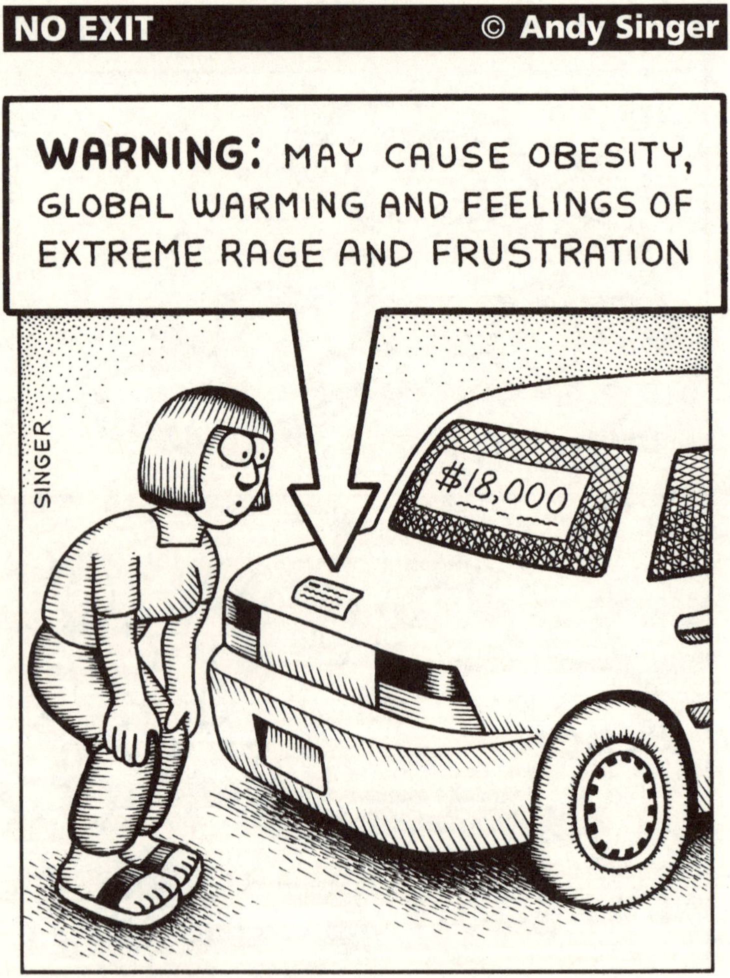
NO EXIT
© Andy Singer
WARNING: MAY CAUSE OBESITY, GLOBAL WARMING AND FEELINGS OF EXTREME RAGE AND FRUSTRATION
SINGER
$18,000

Along the Scenic Route

Harlan Ellison®

The blood-red Mercury with the twin-mounted 7.6 mm Spandaus cut George off as he was shifting lanes. The Merc cut out sharply, three cars behind George, and the driver decked it. The boom of his gas-turbine engine got through George's baffling system without difficulty, like a fist in the ear. The Merc sprayed JP-4 gook and water in a wide fan from its jet nozzle and cut back in, a matter of inches in front of George's Chevy Piranha.

George slapped the selector control on the dash, lighting YOU STUPID BASTARD, WHAT DO YOU THINK YOU'RE DOING and I HOPE YOU CRASH & BURN, YOU SON OF A BITCH. Jessica moaned softly with uncontrolled fear, but George could not hear her: he was screaming obscenities.

George kicked it into Overplunge and depressed the selector button extending the rotating buzzsaws. Dallas razors, they were called, in the repair shoppes. But the crimson Merc pulled away doing an easy 115.

"I'll get you, you beaver-sucker!" he howled.

The Piranha jumped, surged forward. But the Merc was already two dozen car-lengths down the Freeway. Adrenaline pumped through George's system. Beside him, Jessica put a hand on his arm. "Oh, forget it, George; it's just some young snot," she said. Always conciliatory.

"My masculinity's threatened," he murmured, and hunched over the wheel. Jessica looked toward heaven, wishing a bolt of lightning had come from that location many months past, striking Dr. Yasimir directly in his Freud, long before George could have picked up psychiatric justifications for his awful temper.

"Get me Collision Control!" George snarled at her.

Jessica shrugged, as if to say *here we go again,* and dialed CC on the peek. The smiling face of a fusco, the Freeway Sector Control Operator, blurred green and yellow, then came into sharp focus. "Your request, sir?"

"Clearance for duel, Highway 101, northbound."

"Your license number, sir?"

"XUPD 88321," George said. He was scanning the Freeway, keeping the blood-red Mercury in sight, obstinately refusing to stud on the tracking sights.

"Your proposed opponent, sir?"

"Red Mercury GT, '21 model."

"License, sir."

"Just a second." George pressed the stud for the instant replay and the last ten miles rewound on the Sony Backtracker. He ran it forward again till he caught the instant the Merc had passed him, froze the frame, and got the number. "MFCS 90909."

"One moment, sir."

George fretted behind the wheel. "*Now* what the hell's holding her up? Whenever you want service, they've got problems. But boy, when it comes tax time—"

The fusco came back and smiled. "I've checked our master Sector grid, sir, and I find authorization may be permitted, but I am required by law to inform you that your proposed opponent is more heavily armed than yourself."

George licked his lips. "What's he running?"

"Our records indicate 7.6 mm Spandau equipment, bulletproof screens and coded optionals."

George sat silently. His speed dropped. The tachometer fluttered, settled.

"Let him go, George," Jessica said. "You know he'd take you."

Two blotches of anger spread on George's cheeks. "Oh, yeah!?!" He howled at the fusco, "Get me a confirm on that Mercury, fusco!"

She blurred off, and George decked the Piranha: it leaped forward. Jessica sighed with resignation and pulled the drawer out from beneath her bucket. She unfolded the g-suit and began stretching into it. She said nothing, but continued to shake her head.

"We'll *see*!" George said.

"Oh, George, when will you ever grow up?"

He did not answer, but his nostrils flared with barely restrained anger.

The fusco smeared back and said, "Opponent confirms, sir. Freeway Underwriters have already cross-filed you as mutual beneficiaries. Please observe standard traffic regulations, and good luck, sir."

She vanished, and George set the Piranha on sleepwalker as he donned his own g-suit. He overrode the sleeper and was back on manual in moments.

"Now, you stuffer, *now* let's see!" 100. 110. 120.

He was gaining rapidly on the Merc now. As the Chevy hit 120, the mastercomp flashed red and suggested crossover. George punched the selector and the telescoping arms of the buzzsaws retracted into the axles, even as the buzzsaws stopped whirling. In a moment—drawn back in, now merely fancy decorations in the hubcaps. The wheels retracted into the underbody of the Chevy and the air-cushion took over. Now the Chevy skimmed along, two inches above the roadbed of the Freeway.

Ahead, George could see the Merc also crossing over to air-cushion. 120. 135. 150.

"George, this is crazy!" Jessica said, her face in that characteristic shrike expression. "You're no hot-rodder, George. You're a family man, and this is the family car!"

George chuckled nastily. "I've had it with these fuzzfaces. Last year...you remember last year?...you remember when that punk stuffer ran us into the abutment? I swore

I'd never put up with that kind of thing again. Why'd'you think I had all the optionals installed?"

Jessica opened the tambour doors of the glove compartment and slid out the service tray. She unplugged the jar of anti-flash salve and began spreading it on her face and hands. "I *knew* I shouldn't have let you put that laser thing in this car!" George chuckled again. Fuzzfaces, punks, rodders!

George felt the Piranha surge forward, the big reliable Stirling engine recycling the hot air for more and more efficient thrust. Unlike the Merc's inefficient kerosene system, there was no exhaust emission from the nuclear power plant, the external combustion engine almost noiseless, the big radiator tailfin in the rear dissipating the tremendous heat, stabilizing the car as it swooshed along, two inches off the roadbed.

George knew he would catch the blood-red Mercury. Then one smartass punk was going to learn he couldn't flout law and order by running decent citizens off the Freeways!

"Get me my gun," George said.

Jessica shook her head with exasperation, reached under George's bucket, pulled out his drawer and handed him the bulky .45 automatic in its breakaway upside-down shoulder rig. George studded in the sleeper, worked his arms into the rig, tested the oiled leather of the holster, and when he was satisfied, returned the Piranha to manual.

"Oh, God," Jessica said, "John Dillinger rides again."

"Listen!" George shouted, getting more furious with each stupidity she offered. "If you can't be of some help to me, just shut your damned mouth. I'd put you out and come back for you, but I'm in a duel...can you understand that? I'm in a duel!" She murmured a yes, George, and fell silent.

There was a transmission queep from the transceiver. George studded it on. No picture. Just vocal. It had to be the driver of the Mercury, up ahead of them. Beaming directly at one another's antennae, using a tightbeam directional, they could keep in touch: it was a standard trick used by rods to rattle their opponents.

"Hey, Boze, you not really gonna custer me, are you?

Back'm, Boze. No bad trips, true. The kid'll drop back, hang a couple of biggies on ya, just to teach ya a little lesson, letcha swimaway." The voice of the driver was hard, mirthless, the ugly sound of a driver used to being challenged.

"Listen, you young snot," George said, grating his words, trying to sound more menacing than he felt, "I'm going to teach *you* the lesson!"

The Merc's driver laughed raucously.

"Boze, you *de*-mote me, true!"

"And stop calling me a bozo, you lousy little degenerate!"

"Ooooo-weeee, got me a thrasher this time out. Okay, Boze, you be custer an' I'll play arrow. Good shells, baby Boze!"

The finalizing queep sounded, and George gripped the wheel with hands that went knuckle-white. The Merc suddenly shot away from him. He had been steadily gaining, but now as though it had been springloaded, the Mercury burst forward, spraying gook and water on both sides of the forty-foot lanes they were using. "Cut in his afterburner," George snarled. The driver of the Mercury had injected water into the exhaust for added thrust through the jet nozzle. The boom of the Merc's big, noisy engine hit him, and George studded in the rear-mounted propellers to give him more speed. 175. 185. 195.

He was crawling up the line toward the Merc. Gaining, gaining. Jessica pulled out her drawer and unfolded her crash-suit. It went on over the g-suit, and she let George know what she thought of his turning their Sunday Drive into a kamikaze duel.

He told her to stuff it, and did a sleeper, donned his own crash-suit, applied flash salve, and lowered the bang-up helmet onto his head.

Back on manual he crawled, crawled, till he was only fifty yards behind the Mercury, the gas-turbine vehicle sharp in his tinted windshield. "Put on your goggles...I'm going to show that punk who's a bozo..."

He pressed the stud to open the laser louvers. The needle-nosed glass tube peered out from its bay in the Chevy's hood. George read the power drain on his dash. The MHD power generator used to drive the laser was charging. He remembered what the salesman at Chick Williams Chevrolet had told him, pridefully, about the laser gun, when George had inquired about the optional.

Dynamite feature, Mr. Jackson. Absolutely sensational. Works off a magneto hydro dynamic power generator. Latest thing in defense armament. You know, to achieve sufficient potency from a CO_2 laser, you'd need a glass tube a mile long. Well, sir, we both know that's *impractical, to say the least, so the project engineers at Chevy's big Bombay plant developed the "stack" method. Glass rods baffled with mirrors—360 feet of stack, the length of a football field...plus end-zones. Use it three ways. Punch a hole right through their tires at any speed under a hundred and twenty. If they're running a GT, you can put that hole right into the kerosene fuel tank, blow them off the road. Or, if they're running a Stirling, just heat the radiator. When the radiator gets hotter than the engine, the whole works shuts down. Dynamite. Also...and this is with proper CC authorization, you can go straight for the old jugular. Use the beam on the driver. Makes a neat hole. Dynamite!*

"I'll take it," George murmured.

"What did you say?" Jessica asked.

"Nothing."

"George, you're a family man, not a rodder!"

"Stuff it!"

Then he was sorry he'd said it. She meant well. It was simply that...well, a man had to work hard to keep his balls. He looked sidewise at her. Wearing the Armadillo crash-suit, with its overlapping discs of ceramic material, she looked like a ferryflight pilot. The bangup hat hid her face. He wanted to apologize, but the moment had arrived. He locked the laser on the Merc, depressed the fire stud, and a beam of blinding light flashed from the hood of the Piranha. With the Merc on air-cushion, he had gone straight for the fuel tank.

But the Merc suddenly wasn't in front of him. Even as he had fired, the driver had sheered left into the next forty-foot-wide lane, and cut speed drastically. The Merc dropped back past them as the Piranha swooshed ahead.

"He's on my back!" George shouted.

The next moment Spandau slugs tore at the hide of the Chevy. George slapped the studs, and the bulletproof screens went up. But not before pingholes had appeared in the beryllium hide of the Chevy, exposing the boron fiber filaments that gave the car its lightweight maneuverability. "Stuffer!" George breathed, terribly frightened. The driver was on his back, could ride him into the ground.

He swerved, dropping flaps and skimming the Piranha back and forth in wide arcs, across the two lanes. The Merc hung on. The Spandaus chattered heavily. The screens would hold, but what else was the driver running? What were the "coded optionals" the CC fusco had mentioned?

"Now see what you've gotten us into!"

"Jess, shut up, shut up!"

The transceiver queeped. He studded it on, still swerving. This time the driver of the Merc was sending via microwave video. The face blurred in.

He was a young boy. In his teens. Acne.

"Punk! Stinking punk!" George screamed, trying to swerve, drop back, accelerate. Nothing. The blood-red Merc hung on his tailfin, pounding at him. If one of those bullets struck the radiator tailfin, ricocheted, pierced to the engine, got through the lead shielding around the reactor. Jessica was crying, huddled inside her Armadillo.

He was silently glad she was in the g-suit. He would try something illegal in a moment.

"Hey, Boze. What's your slit look like? Is she nice? If she's creamy'n'nice I might letcha drop her at the next getty, and come back for her later. With your insurance, baby, and my pickle, I can keep her creamy'n'nice."

"Fuzzfaced punk! I'll see you dead first!"

"You're a real thrasher, old dad. Wish you well, but it's soon over. Say bye-bye to the nice rodder. You gonna die, old dad!"

George was shrieking inarticulately.

The boy laughed wildly. He was up on something. Ferro-coke, perhaps. Or D4. Or merryloo. His eyes glistened blue and young and deadly as a snake.

"Just wanted you to know the name of your piledriver, old dad. *You* can call me Billy..."

And he was gone. The Merc slipped forward, closer, and George had only a moment to realize that this Billy could not possibly have the money to equip his car with a laser, and that was a godsend. But the Spandaus were hacking away at the bulletproof screens. They weren't meant for extended punishment like this. Damn that Detroit iron!

He had to make the illegal move *now*.

Thank God for the g-suits. A tight turn, across the lanes, in direct contravention of the authorization. And in a tight turn, without the g-suits, doing—he checked the speedometer and tach—250 mph, the blood slams up against one side of the body. The g-suits would squeeze the side of the body where the blood tried to pool up. They would live. If...

He spun the wheel hard, slamming down on the accelerator. The Merc slewed sidewise and caught the turn. He never had a chance. He pulled out of the illegal turn, and their positions were the same. But the Merc had dropped back several car-lengths. Then from the transceiver there was a queep and he did not even stud in as the Police Copter overhead tightbeamed him in an authoritative voice:

"XUPD 88321. Warning! You will be in contravention of your dueling authorization if you try another maneuver of that sort! You are warned to keep to your lanes and the standard rules of road courtesy!"

Then it was queeped, and George felt the universe settling like silt over him. He was being killed by the system.

He'd have to eject. The seat would save him and Jessica. He tried to tell her, but she had fainted.

How did I get into this? he pleaded with himself. *Dear God, I swear if you get me out of this alive I'll never never never go crazy like this again. Please God.*

Then the Merc was up on him again, pulling *alongside*!

The window went down on the passenger side of the Mercury, and George whipped a glance across to see Billy with his lips skinned back from his teeth under the windblast and acceleration, aiming a .45 at him. Barely thinking, George studded the bumpers.

The super-conducting magnetic bumpers took hold, sucked Billy into his magnetic field, and they collided with a crash that shook the .45 out of the rodder's hand. In the instant of collision, George realized he had made his chance, and dropped back. In a moment he was riding the Merc's tail again.

Naked barbarism took hold. He wanted to kill now. Not crash the other, not wound the other, not stop the other—*kill the other!* Messages to God were forgotten.

He locked-in the laser and aimed for the windshield bubble. His sights caught the rear of the bubble, fastened to the outline of Billy's head, and George fired.

As the bolt of light struck the bubble, a black spot appeared, and remained for the seconds the laser touched. When the light cut off, the black spot vanished. George cursed, screamed, cried, in fear and helplessness.

The Merc was equipped with a frequency-sensitive laser-proof windshield. Chemicals in the windshield would "go black," opaque at certain frequencies, momentarily, anywhere a laser light touched them. He should have known. A duelist like this Billy, trained in weaponry, equipped for whatever might chance down a Freeway. Another coded optional. George found he was crying, piteously, within the cavern of his bangup hat.

Then the Merc was swerving again, executing a roll and dip that George could not understand, could not predict. Then the Merc dropped speed suddenly, and George found himself almost running up the jet nozzle of the blood-red vehicle.

He spun out and around, and Billy was behind him once more, closing in for the kill. He sent the propellers to full spin and reached for eternity. 270. 280. 290.

Then he heard the sizzling, and jerked his head around to see the back wall of the car rippling. *Oh my God,* he thought, in terror, *he can't afford a laser, but he's got an inductor beam!*

The beam was setting up strong local eddy currents in the beryllium hide of the Chevy. He'd rip a hole in the skin, the air would whip through, the car would go out of control.

George knew he was dead.

And Jessica.

And all because of this punk, this rodder fuzzface.

The Merc closed in confidently.

George thought wildly. There was no time for anything but the blind plunging panic of random thought. The speedometer and the tach agreed. They were doing 300 mph.

Riding on air-cushions.

The thought slipped through his panic.

It was the only possibility. He ripped off his bangup hat, and fumbled Jessica's loose. He hugged them in his lap with his free hand, and managed to stud down the window on the driver's side. Instantly, a blast of wind and accelerated air skinned back his lips, plastered his cheeks hollowly, made a death's head of Jessica's features. He fought to keep the Chevy stable, gyro'd.

Then, holding the bangup hats by their straps, he forced them around the edge of the window where the force of his speed jammed them against the side of the Chevy. Then he let go. And studded up the window. And braked sharply.

The bulky bangup hats dropped away, hit the roadbed, rolled directly into the path of the Merc. They disappeared underneath the blood-red car, and instantly the vehicle hit the Freeway. George swerved out of the way, dropping speed quickly.

The Merc hit with a crash, bounced, hit again, bounced and hit, bounced and hit. As it went past the Piranha, George saw Billy caroming off the insides of the car.

He watched the vehicle skid, wheelless, for a quarter of a mile down the Freeway before it caught the inner breakwall of the Jersey Barrier, shot high in the air, and came down turning over. It landed on the bubble, which burst, and exploded in a flash of fire and smoke that rocked the Chevy.

At three hundred miles per hour, two inches above the Freeway, riding on air, anything that broke up the air bubble would be a lethal weapon. He had won the duel. That

Billy was dead.

George pulled in at the next getty, and sat in the lot. Jessica came around finally. He was slumped over the wheel, shaking, unable to speak.

She looked over at him, then reached out a trembling hand to touch his shoulder. He jumped at the infinitesimal pressure, felt through the g- and crash-suits. She started to speak, but the peek queeped, and she studded it on.

"Sector Control, sir." The fusco smiled.

He did not look up.

"Congratulations, sir. Despite one possible infraction, your duel has been logged as legal and binding. You'll be pleased to know that the occupant of the car you challenged was rated number one in the entire Central and Western Freeway circuits. Now that Mr. Bonney has been finalized, we are entering your name on the dueling records. Underwriters have asked us to inform you that a check will be in the mails to you within twenty-four hours.

"Again, sir, congratulations."

The peek went dead, and George tried to focus on the parking lot of the neon-and-silver getty. It had been a terrible experience. He never wanted to use a car that way again. It had been some other George, certainly not him.

"I'm a family man," he repeated Jessica's words. "And this is just a family car...I..."

She was smiling gently at him. Then they were in each other's arms, and he was crying, and she was saying that's all right, George, you had to do it, it's all right.

And the peek queeped.

She studded it on and the face of the fusco smiled back at her. "Congratulations, sir, you'll be pleased to know that Sector Control already has fifteen duel challenges for you.

"Mr. Ronnie Lee Hauptman of Dallas has asked for first challenge, and is, at this moment, speeding toward you with an ETA of 6:15 this evening. In the event Mr. Hauptman does not survive, you have waiting challenges from Mr. Fred Bull of Chatsworth, California...Mr. Leo Fowler of Philadelphia...Mr. Emil Zalenko of..."

George did not hear the list. He was trying desperately,

with clubbed fingers, to extricate himself from the strangling folds of g- and crash-suits. But he knew it was no good. He would have to fight.

In the world of the Freeway, there was no place for a walking man.

2003 Postscript

When I arrived in Los Angeles on New Year's Day 1962, I discovered a city more wonderful than I had ever imagined. It had all the verve and passion of New York during its last really vibrant period—'54-'57—coupled with the insouciance of New Orleans before it became aware of itself as The Big Easy. It had the charm of small-town America and the urbanity of Big Town USA.

All that is gone, today. The demon of "Progress" as worshipped by ex-Governor Reagan and all the clone-idiots who have followed him in office, has turned LA into yet another sprawling, choking, murderous, inconvenient, imbecile paradise for realtors, rampant builders, illiterate gangs of slaughtering assholes, weirdos of every stripe, and that great American icon, OJ Simpson, may his dick fall off in his soup.

Nonetheless, I am an Angeleno. I've lived here longer than anywhere else in my peripatetic life, and I cannot think of a better place to reside, for all its drawbacks. (No, I'm not nuts about the earthquakes, but you've got the equivalent crap where you are—twister, flood, smog, Fundamentalists —so that part's a tradeoff.)

I wrote this story at least ten years, maybe more, before actual driveby shootings became common throughout the U.S., but commencing with LA at the cutting edge. I wrote it on the subject of middle-class conformity, mediocrity, and the utter stupidity of *Macho* behavior. The idea that such a wild and improbable social escapade could become commonplace in less than a decade, never occured to me. When the first freeway "duels" began out here, I was contacted by the AP, Reuters, the UP, the INS, and a dozen newspapers. How they had made the liaison between my little fable and the new reality, I never found out. Maybe someone working the *NY Times* night desk had read the story. They all asked me how I felt about "predicting" this phenomenon, and wasn't this another classic example of how insightful "sci-fi" was supposed to be? I didn't have the heart to tell them that if you've got hundreds of writers all wildly predicting anything they can think of, that eventually there will be a few hits (but in fact, sf has as lousy a hit-to-miss ratio on prediction and extrapolation as the most bogus fortune-teller found in the tabloids). Didn't have the heart—particularly after they pointed out that I'd coined the term "fax" in another story—and decided to accept the mantle of Nostradamus.

All bullshit, of course, but how else are we clowns to stay in business? You can fool some of the people some of the time, and some of the people some of the time, but you can't fool some of the people some of the time. I think I've got that right.

Buscrunch

Dean Wirth

Bus crunch thanks a bunch, Maybe I be dead by lunch
Car door give me more, Crushed kneecap what a bore
Niquil Refill, Fist full of Waybills
Couriers are evil, Weaving through traffic like snakes
Oh, look it stopped raining, Maybe now I'll have brakes
Accident, get me bent, Hospital time is time well spent
Repeat this is not a test, It's a blood soaked manifest
Loving cup, muck me up, Dragged on asphalt by a pickup truck
How now puppy chow, Just don't laugh you're not allowed
Stone breath nothing left, Look I'm being reg'd to death
Slop cop make me stop, Keep me from my next drop
Bumper ride things collide, Meet you on the other side
Trips for money money for trips, Yes I'm just doing this for kicks
Ambulance siren dance, You just blew your second chance
Rain hail snow sleet, Just a piece of courier meat
Dead now take a bow, It's okay to draw a crowd

Documentation: Portland's Memorial Lifehouse

Peter Gelman

With an illustration by Keith McKellar

One day in May of 1998, in Portland, Oregon, Matthew Schekel died on his way to work. He was bicycling across a residential intersection of SE 37th Avenue and Taylor Street, near the popular Hawthorne District. A delivery truck rolled through the stop sign and killed him.

Soon afterward, someone locked the broken bike to the stop sign. A cross stood with it, along with flowers, poems, and personal artifacts. Over the next four years, many stopped to examine this sad display. Noticing how the neighborhood came to cherish the shrine for the fallen cyclist, the neighborhood organization, Southeast Uplift, worked with another group, City Repair, to build a formal structure as part of a workshop of natural builders.

City Repair helps communities create public space while promoting sustainability and livability. 'Livability' is a buzzword that many use when discussing Portland's relatively unique direction as a U.S. city, with its thirty year-old urban growth boundary against sprawl, for example. In this case of this shrine, the word 'livability' takes a double meaning and urgency.

Chiefly, the "Memorial Lifehouse" is an eight-foot column that looks a bit like a lighthouse, as the name suggests. The lower half of the column consists of masonry; the upper half

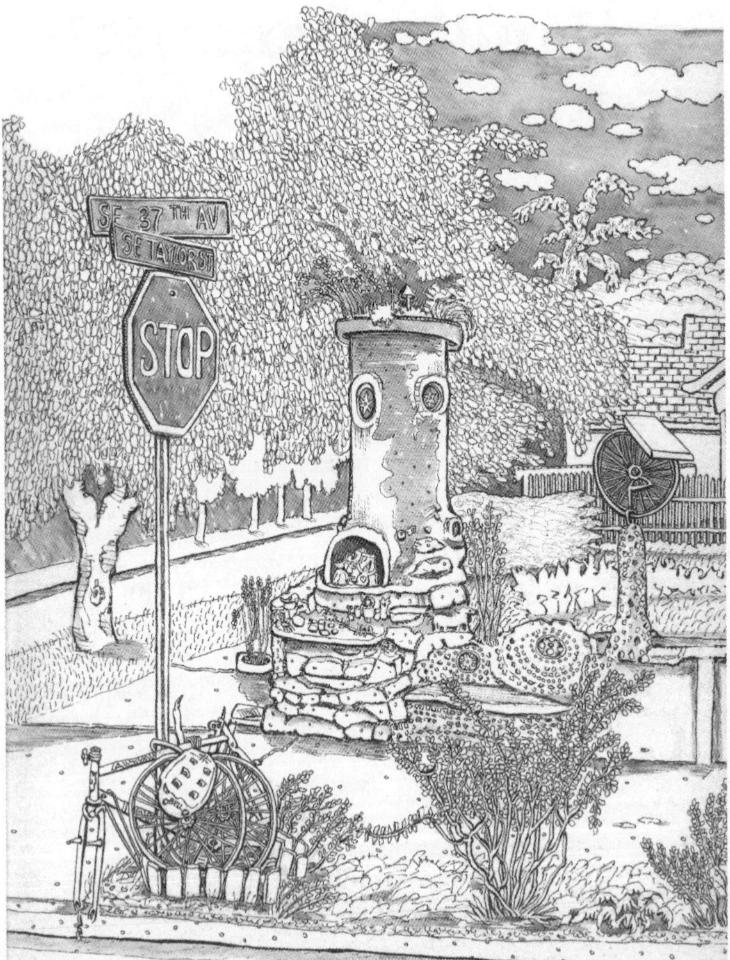

is "cob" (a building material made of earth, sand and straw). Stained-glass bike wheel-windows at top glow at night under a peaked "green roof"—a roof on which plants grow.

Next to the column there is a seat with a mosaic, including bicycle parts and other artifacts. On the seat's other side, a small concrete column mounts a bicycle fork that holds a solar panel. The panel powers the lantern through a battery in the main tower. In the back, an inflated tricycle wheel helps seal the maintenance hatch. The site also includes an explanatory sign on a painted bike wheel frame.

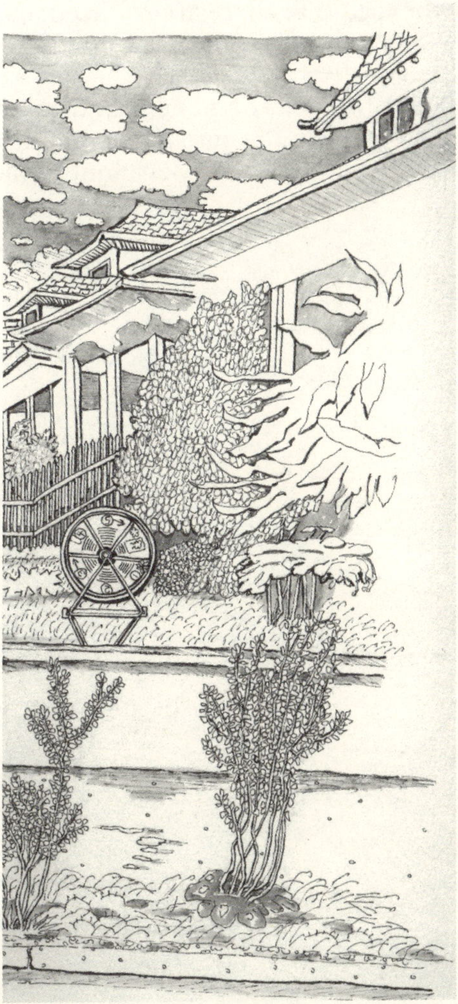

The natural builders met in May, 2002, to foster excitement about environmental building techniques, especially cob. Volunteers built at five public sites, one of which was the Lifehouse. To support the environmental construction, Portland's Office of Sustainable Development donated $7,000. The Bicycle Transportation Association helped organize volunteers for the memorial project.

A natural building teacher from

"Builders without Borders" led the work.

In its first eighteen months of existence, the Memorial has succeeded as a shrine and more. It has added beauty and interest to the intersection. It has slowed down automobile traffic in the intersection. It has become a "place" in the neighborhood, a focal point for cyclists to rally and put up signs. Local newspapers and TV have featured it. For those inclined to think about such things, the Memorial has become an advocate for the idea that there are connections between safety, livability, public space, natural building, and bicyclists.

The following year (2003) several Portland neighborhoods worked with City Repair to create more public spaces, including one beside a cooperative bike shop on Southeast Ankeny Street.

Unemployment is Rising

Grace Tierney

My name is Angela and I started to work for the Boss about ten years ago. She had worked hard for years before that in order to afford my services and initially at least the job was perfect.

It's quite rare in my line of work to get close to your boss but I did. Ah yes, we did. I was at work almost seven days a week and I loved it. We went everywhere together, town for shopping, university for her post-graduate studies, deliveries to the recycling center, visits to her friends. Hey we even went to have mud packs and rainwater treatments together and aromatic oil therapy was part of my health-plan. It was a *great* job and I did it well. She never had to pay for parking, and with me around she never had to wait for a bus or train.

Of course it was hard work too, often with unsocial hours and always carrying the health dangers that go with this kind of work. But I'm an old hand now and the larger pushy vehicles on the roads don't scare me these days. You know, trucks with badly tied loads, buses marking un-signalled turns and taxicabs stopping dead on junctions to collect passengers. It's all just the everyday life of traffic on our streets. I may be getting a bit creaky in the limbs these days but those facts won't put me off the roads, even after losing some old friends to such incidents recently.

Until last year I doubt you could have found a happier employee. The work is rewarding, the benefits were top-notch and the Boss never pushed me too hard. But then He came along. It's always a man, isn't it? He was pleasant enough initially. He even gave me a lift home to my tiny cottage after a long work-day. But gradually it dawned on me, I was barely seeing her anymore. If she needed to get groceries he would drive her there. If she was going out for the night he would accompany her in a taxicab. He was coming between us and it broke my heart.

Things got serious when they started the weekends away. Suddenly I was never needed at weekends because they were off petrol-burning their merry way across the countryside. She started to tell me about how free it felt to throw a bag in the car (this was said with a pitying look at my own luggage capacity) and just "take off on the open road".

After that I wasn't surprised when they moved in together. My role was gradually diminished and my job-linked cottage was downgraded to a wet and draughty tent! I could hardly believe it. The final straw, however, was the day I saw her with her driving license application form, all filled in and ready to submit. I knew then that I was about to become unemployed.

I tried to put my spoke in about it. I tried to squeak a protest but the rust had already taken hold. This will probably be the last year I will survive the neglect under the tarpaulin in the back garden. But hey, that's life in the cycle lane for you.

Henry Ford's Bicycle

Michael Burton

"Henry Ford's bicycle,"

I overheard someone say last night,
while waiting in line at Lounge Ax.

Which makes me wonder:
Could one bicycle have saved us from him?

Like the red Raleigh
that provided escape
from my awkward adolescence.

Like the purple Schwinn
that delivered me
from the fucked-up job and crumbling relationship
to a summer of European back-roads.

Like the grey Trek
that everyday liberates
my lethargic body and mind from desk-job doldrums,
providing street level inspiration
for a struggle of mighty purpose.

Henry Ford's bicycle
did not him take to a higher place,

or provide escape from the demons that haunted his soul.

The demons
destroyed our cities;
isolated our brothers and sisters in mobile iron cages;
and created a nation of flaccid TV watchers,
lulled into complacency by car commercials.

May our bicycles
prove worthy
where his was lacking.

Documentation: Declaration of the Right to Walk and Roll

WALK Austin

Preface
By Reuben Leslie Jr.

The Declaration of Sentiments, modeled after the Declaration of Independence of the United States of America, was written and adopted at the first Women's Rights Convention in the United States, held in Seneca Falls, New York, the nineteenth and twentieth of July, 1848. After initial derision, many setbacks, and a long battle, women's right to vote became a reality in 1920. Like the Magna Carta and the Declaration of Independence, the Declaration of Sentiments did not create the rights it articulated; it affirmed the rights that we believe we are born with in the kind of society in which we wish to live. Consciousness of these rights and the need to express them arose because of abuses that grew intolerable over time. In the United States, the compelling logic and clear expression of inherent liberties in the Declaration of Independence laid the groundwork in social and political thought for the emancipation of slaves and women's suffrage, concepts the signers may not have envisioned. The Declaration of Sentiments was condemned by many; it mimicked the Declaration of Independence and in so doing was seen to trivialize a revered past social and political revolution. In hindsight, nearly everyone in the U.S. now accepts and takes for granted the truth and right of what both documents say.

Inspired by this example of what can be accomplished—monumental progress in social consciousness through articulating and defending a right not previously committed to writing and using the document to overcome opposition by itemizing the abuses of a little respected right and appealing to the profoundly convincing core belief in human freedom and dignity...

IN AUSTIN, TEXAS, MARCH 22, 1997
We, the members of WALK Austin do
now unanimously make this

Declaration of the Right to Walk and Roll

When in the course of human events, it becomes necessary for one portion of the family of humankind to assume among the people of the earth a position different from that which they have hitherto occupied in recent memory, but one to which the laws of nature and of nature's God entitle them, a decent respect to the opinions of others requires that they should declare the causes that impel them to such a course.

We hold these truths to be self-evident: that all human beings are created equal; that they are endowed by their Creator with certain inalienable rights; that among these are life, liberty, and the pursuit of happiness; that to secure these rights governments are instituted, deriving their just powers from the consent of the governed. Whenever any form of government becomes destructive of these ends, it is the right of those who suffer from it to refuse allegiance to it, and to insist upon the institution of a new government, laying its foundation on such principles, and organizing its powers in such form as to them shall seem most likely to effect their safety and happiness. Prudence, indeed, will dictate that governments long established should not be changed for light and transient causes; and accordingly, all experience has shown that humankind are more disposed to suffer, while evils are sufferable, than to right themselves, by abolishing the forms to which they are accustomed. But when a long train of abuses and usurpations, pursuing invariably the same object, evinces a design to reduce them under absolute despotism, it is their duty to reform or throw off such government, and to provide new safeguards for their future security. Such has been the patient sufferance under this government of the men, women, boys and girls not in motor vehicles, and such is now the necessity which constrains them to demand the equal station to which they are entitled.

The history of city-building since the takeover by automobiles earlier in this century is a history of repeated injuries and usurpations on the part of a partly callous, partly thoughtless, partly shortsightedly "modernist" controlling majority in positions of power in government, design, construction, and industry, particularly automotive manufacturing and retail, and the press, toward people in cities who are not in motor vehicles, having in direct object the establishment and proliferation of an absolute proscription from streets and roadways of people inconvenient to speeding motor vehicles. To prove this, let facts be submitted to a candid world:

> This controlling majority has never acknowledged or permitted pedestrians to exercise the inalienable right to walk and roll in safety and comfort along and across all streets.
>
> It has compelled pedestrians to submit to laws and designs, in the formation of which they have

not been considered equals to people driving vehicles, whose turning radius, speed, and freedom from even minor delays have been considered paramount.

It has withheld from them the right to access to whole sections of the city carved away and separated by wide interchanges that ban pedestrians or which are instead given generously to the visitors foreign and domestic from suburb, farm, or other cities, near and far, passing through in vehicles or parking their vehicles and visiting briefly, ironically valuing their pedestrian experiences in a few special places as highlights of visits and even preferred destinations, unmindful of the local communities nearby or through which these visitors travel only in motor vehicles.

Having deprived them of this first right of an urban citizen, the right to walk and roll on all streets, thereby leaving them without safe transportation, and without equal representation in the halls of the market place or the legislature, it has oppressed them on all sides.

It has made them, if walking or rolling along streets without sidewalks, in the eye of the law, hazards and violators of motorists' rights; their peace, comfort, safety, health, and lives are held in such low esteem that punishments for violating their rights and liability to them or their estates for wrongs are seldom seriously pursued.

It has taken from them all practical right to public right of way, even to travel to and from jobs for wages enough to buy an automobile, or to travel to an automobile dealer, or visit a cemetery.

It has made them, morally, an irresponsible being, as they must commit many crimes, with impunity, provided they be done out of the presence of motor vehicles. In the intersection, they are compelled to yield to any vehicle, it becoming, to all intents and purposes, their master—the law giving it power to deprive them of their lives and liberty, and to administer chastisements of loud motor-revving, tire-screeching, horn-honking, shouts, foul smells, insulting gestures, littering, harsh wind gusts, dust-gravel-mud-or-water spraying, vehicular assault, and injury or death.

It has so framed the laws of traffic and development as to what shall be the proper causes of citations and variances; in case of public right-of-way for whom the use of public easement for walking by children or adults shall be given, as to be wholly regardless of the happiness of pedestrians—the law, in all cases, going upon the false supposition of the supremacy of vehicle users and property owners disrespectful of community, and giving all power into their hands.

After depriving them of all rights as an equal user of public infrastructure, if the owner of property or renter whose landlord pays taxes from rents, it has taxed them to support a government which recognizes them only as taxpayers and hazards to protect motorists from with expensive and ever more complex interchanges, wider lanes and buffers, higher fences, and faster-flowing, uncrossable roadways.

It has allowed motor vehicles to monopolize and destroy the land through pollution and erosion, and to fill the land with roads and parking lots and otherwise despoil from pedestrian view and use nearly all the beautiful and convenient thoroughfares between homes and profitable employments. And from too many of such areas that they are permitted access to they can receive but a scanty use because of short-timed or missing traffic lights; absence of drinking, washing or rest room facilities; broken, automobile-blocked, garbage-container-blocked, littered, automatic-sprinkler sprayed, boulder-or-brush-covered, low-tree-limb canopied; or absent sidewalks; or sidewalks with ill-placed mail-boxes, sign and utility poles, utility boxes, vehicle barricades, construction materials and equipment, and other hazards. For any access without a personal vehicle to most areas they must endure great hardship to reach public transit services because of the long distances resulting from motor-vehicle-scaled, low-density and segregated-use development.

It closes against them the avenues landscaped with greatest beauty, wealth, distinction and prestige, which it considers most honorable for motorists to view fleetingly from a distance as they speed by.

It has denied them and itself the facilities and government-supported institutional means of receiving and creatively adapting—to make more socially inclusive and democratic as well as more technologically sophisticated—the rich cultural and practical scientific traditions, education, appreciation, understanding and consciousness of the inter-related humane, aesthetic, ethical, sensual, spiritual, and intellectual theory and practice of critical aspects of important human activity including but not limited to playing, resting, socializing, talking, watching, walking, shopping, selling, cooking, sitting, contemplating, as well as the disciplines of economics, planning, administration, sociology, language, accounting, child development, gerontology, medicine, law, religions, commerce, geography, politics, anthropology, psychology, natural sciences, journalism, history, philosophy, literature, design, architecture, performing and visual arts, crafts and not least engineering, that are essential to creating and enjoying the infrastructure required for a free, democratic and civilized community in a modern urban environment—streets for people.

It allows them access to shopping, worship, business, dining, entertainment, schools and government buildings but only through inconvenient subordinate means, claiming design aesthetics (such as broad lawns and parking lots dotted with exotic plants suited to golf-courses or hothouses) and demands of safety and efficiency of traffic flow as authority for their exclusion from many roadways and entrances off parking lots, resulting in long unsheltered walks from transit stops to building entrances or along roadways.

It has created a false public sentiment, by giving to the world a different code of morals for them and people in vehicles and by which moral delinquencies which exclude from streets and endanger pedestrians are not only tolerated but deemed of little account.

It has usurped the prerogative of God, claiming it as its right to assign for them a sphere of action, when that belongs to their consciences and their God.

It has endeavored, in every way that it could to destroy the confidence of people who walk in their own powers, to lessen their self-respect, and to make them willing to lead a polluting, expensive, isolated, vehicle-dependent and abject life or walk and roll in misery, danger and fear.

Now, in view of this entire disfranchisement of—temporarily at least at one time or another—all the people of the city, their social and physical degradation, in view of the unjust laws above mentioned, and because pedestrians do feel themselves aggrieved, oppressed, and fraudulently deprived of their most sacred rights, we insist that they have full, immediate admission to all the rights and privileges which belong to them as human beings and as citizens of these

United States, the State of Texas, the County of Travis, and the City of Austin.

In entering upon the great work before us, we anticipate no small amount of misconception, misrepresentation, and ridicule; but we shall use every instrumentality within our power to effect our object. We shall employ agents, circulate tracts, petition the city, county, regional, state and national legislatures, and endeavor to enlist the pulpit, performing and visual arts and all communications media in our behalf. We hope this event will be followed by a series of like events, embracing every part of the city, and eventually the county, the state, and the country.

Firmly relying upon the final triumph of the Right and the True, we do this day affix our signatures to this declaration.

On Driving: Three Poems

Patricia Wellingham-Jones

Flatlanders Head for the Hills

The '51 brand new car cranked around hair-pin turns,
swooped through crossings, out-ran stray dogs.
Father clutched the wheel, blue eyes blazing.
Mother, pale beside him, clucked her tongue.
Bottoms bounced, rough-fibered seat
rubbed red patches on tan bare skin.

My sister and I peered through open windows'
rushing air at ambling cows, valley lakes and peaks.
Shrilled our excitement
in Father's overwrought ear.
Mother wrung thin hands, moaned soft and low.

We roared through Adirondacks
to our cousin's country home,
left a few feathers in our wake.
Final flourish of the wheel,
brakes shrieked inches from the door.

Father swaggered, Mother staggered from the car.
Amply impressed, my uncle asked,

When did you get your license, Norm?
Father beamed in modest pride,
I've had it one whole week.

Highway Signs

Stark
the white sticks
crossed high
just like the bones
they represent
from rock piles
or small graves
by the roadside.

Plastic
flowers loop
in drunken abandon,
reminders
of the absent
drivers they replace
or huddle
posey-like
in brown string.

Ribbons
blow to tatters
in the wind,
red
as blood pooled
on the highway,
white
as visible prayer.

Over a hill
boom box blasting,
pickup
lights flash
across the crossed sticks,

disappear
in waves of laughter,
tossed cans.

The night
bides its time.

Important

He talks extra loud
in restaurants,
paces city streets in a daze.
Sidesweeps mailboxes,
sails through stop signs,
tailgates trucks.
Through daily life
he skims unheeding,
cell phone clapped to his ear.

This Machine Cures Melancholia

Peter Gelman

With an illustration by Neal Skorpen

As it didn't rain, the January sky didn't need clouds: that stranger, Yellow Face, peaked down from the top of Mount Hood. Jake, Antonia and I agreed to bike toward the Columbia River on the path that traced highway 205. "I put in new brake pads, finally," Jake told me unhappily. "The kind you recommended, with the little tooth that dries the rim."

I nodded cautiously, suspecting that Jake and Antonia had been bickering before I arrived in my own bad mood. "In about five minutes we'll all be happy," I told them. Antonia half-smiled because I always say that when we start out; then she was racing past me.

I caught up, and told her: "Think about what Spanish word you want to teach me today, okay?"

"Okay!" she nodded, preparing to begin the lesson, but I interrupted—"but not now, wait twelve and a half minutes"—"Why twelve and a half minutes?"—"Because twelve and a half minutes on a bike puts anyone into a good mood."

She scoffed, "It will take at least twenty minutes today."

I pulled ahead to leave her and Jake to themselves awhile. Somewhat grimly, we started off uphill and as my effort increased I felt my mood lighten. I felt the clenched unhappiness of my mind drop away and felt the gratification of altitude for my efforts.

As we passed Fred Meyers uphill I realized I had pulled far ahead. Jake and Antonia had stopped behind me, so I waited. I yelled back, "Sorry didn't see you stop!"

They started to cycle up the hill. Jake yelled "Yabba babba babba!" or rather, that was what it sounded like through the wind.

So I yelled back, "Babba yabba yabba!"

Despite his low speed, Jake managed to skid to a stop beside me. He asked, "Hey, did you actually say 'Babba yabba yabba'?"

"Yes siree, but I was just repeating back what I heard."

As Antonia was catching up to us, Jake explained what he actually said: Antonia had stopped because she saw three pennies by the curb and she remembered dreaming the night before that she found three pennies.

I looked at her as she stopped beside us. "You want us to wait while you go spend your three pennies at Fred Meyers?"

Familiar with my monkey chatter, she rolled her eyes. "With three pennies?"

"I know something you can buy for three cents," Jake smirked.

"What?"

We waited. Jake turned away and quickly peddled away from us without answering.

Still we waited. Suddenly he turned over his shoulder and shouted back gleefully, "You can buy some yabba babba babba!"

As we neared the highway the traffic noise grew louder. We turned left at the Gateway Transit Center. A lanky runner with an elf cap passed us.

"Hey you can see they made a lot of progress on the light rail line to the airport," I told them through the noise of a bus pulling nearby.

"They made a lot of yabba babba babba," Jake agreed.

I immediately retorted, "Don't you yabba babba babba me. And right here is where that security guard told me to walk my bike. He was polite but he looked like he wanted to kick my butt."

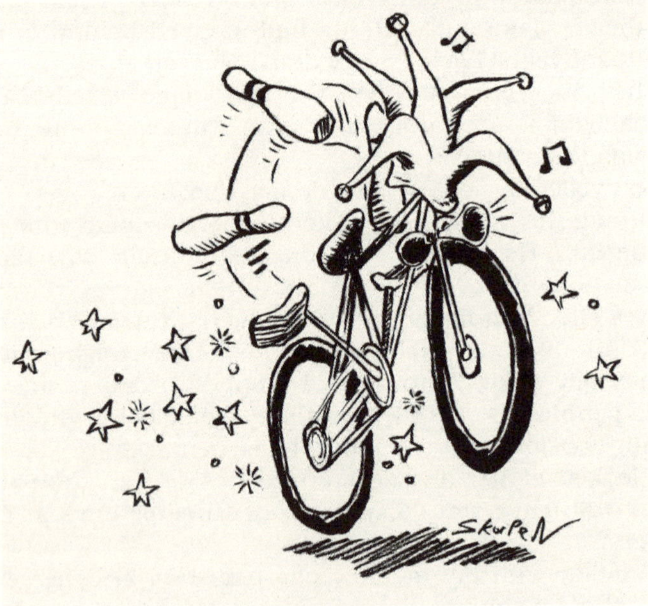

"Oh, the wind," Antonia shivered.

The wind flapped and fluttered but the sun still shone. The traffic noise roared nearby. We rolled downhill quickly and passed the runner in the elf cap. In a few minutes we rose up again near Rocky Butte, the noisy highway between us. The path curved further uphill and then we saw Mount Saint Helens, so white the white was blue, with pink shadows, in the winter sun.

"Wow! See that?" grinned Jake.

"Ha! I told you!"

"What?"

"We're in good moods now."

The runner in the elf cap suddenly came up from behind us and passed us.

"Not just good moods. Great moods!" Antonia smiled, shivering. "Look at Hood."

It stood totally white, like a canine tooth, in the east.

"In Mexico, you name your mountains names like Popotenochtectopopot..." I began.

"You mean Popacatpetl?"

"Yeah, that's what I said, and we call our mountain Hood. Just Hood. One syllable."

"Funny," Jake observed. "A one-syllable word seems absurdly inadequate to name a mountain. I'm not sure a five or six syllable word is any better."

Antonia did not seem surprised. "Maybe all names for mountains are absurd."

"Maybe we need a hundred syllable word," I suggested helpfully.

Faintly, Antonia smiled. "They have no names, really." She suddenly rang her bicycle bell. When I looked at her, she asked, "Are you ready for your word?"

I nodded as I steered the bike around the broken red plastic shards of a car's brake light.

"Spanish for 'flat tire' is *llanta ponchada.*"

I stood up on the pedals and called out robotically: "Violation. Violation! I detect a violation! No fair, it's two words. I'm only supposed to learn one!"

"Too bad! Now you say it: *llanta ponchada.*"

"What? something *thonthatha*?"

"*Llanta* with two 'l's that sound like an English 'y'. *Llanta* like *llama.*" Patiently, she repeated, "*Llanta ponchada.*"

I decided not to be cooperative. "Oh, two 'l's that sounds like an English 'why'. I get it. *Why-yabba babba babba.*"

"No! Very bad. Are you trying to be stupid?"

"Yes, it's my natural talent."

It took a half a mile to admit to her that I could say it right and then she taught me how to spell it.

We passed the elf-cap runner and rolled on North with the airplanes buzzing overhead. We passed a hotel with ESPN where my friends held a Tour de France party. We passed Jim Dandies, a shabby little bar where we went to drink afterwards—it was great fun. About nine miles from

Jake and Antonia's house we went on toward the highway 205 bridge which has a bike lane over to Washington State where I crashed once, stupidly, and broke my rear rack. It amused me to think of how I could look left, look right, and with each look trigger a significant cycling memory. The clouds closed and the Gorge wind picked up but I was warm from the cycling.

Suddenly I heard behind me the sound of a glass thermos exploding. That was Jake's tire. As we began to take apart the tire and tube we saw Mexican cactus thorns stuck in many places in the tread, poking into the tube. But the tube was full of anti-flat glue so all this time, a year since they left Mexico, it held. The thorns were small but tough, hard to pull out; Jake struggled with the tire.

I squatted down to help fix the tube but the blue glue-liquid spilled out on my wind jacket. It would destroy the expensive fabric wherever it touched: I sprang up, dodging the flying blue drops. How could I patch that wet gooey tire without becoming gooey myself?

"I have a new tube in my seat pack," Jake muttered as he unzipped it; I threw the wet tire down.

Meanwhile I started feeling the cold grip me as I stood there in my bike shorts in that winter wind. With the cold came a bleak feeling; the grass by the path was yellow-grey. I realized the small toe of each foot was numb; I stamped my feet. The elf-cap runner passed us once again: he was the tortoise and we the hares.

Antonia, meanwhile, was investigating the cause of the damage. Antonia was examining the angle of the brake pads. "This brake pad doesn't have a tooth," she exclaimed. "It's a fang. You put it at a bad angle."

Jake set down the patched tube. "I did?"

"See, it rubbed on the side of the tire and chewed it open."

He just lifted his eyebrows and grunted. Now he was poking a finger through the tear in the sidewall of the tire. Jake shrugged, and started to put the tube and tire on the rim.

I whistled. "Whoa, there Jake. Need a tire boot?"

He looked concerned. "What's a tire boot?"

"A heavy-duty patch for a tire. I learned to carry one because once I didn't have one and I tried patching the tire with a tube patch, which burst three blocks later." I gave Jake the tire boot from my under-seat bag and I told them, "Here's a helpful hint from all the bike magazines. If you put a gash in your tire and don't have a boot, pull out your wallet. Cover the gash in your tire with a dollar bill."

"A dollar bill?"

"Yeah, put it between the tire and the tube. Use a twenty, if you prefer. American money is superpower money and it can do anything, right? Not counting universal health care or full employment."

"Yeah," Jake grunted, looking at the tire boot. "I guess money's made to survive the grip of a million fingers." He quickly put the wheel in the stays and with his folding tool, repositioned the brake pads.

Antonia, Jake and I started back now, all of us shivering. It was better to be moving. In the motion I felt freedom from the past and in the effort I felt a spontaneity and continuous reward. The past, like every foot I zipped over, was insubstantial—the bike, the road beneath, the engine of my body and the mobile fellowship of my companions—this was real. I didn't brood over the return of the gray sky or the traffic as cycling always puts me in a good mood and I warmed up again soon. With a tailwind and long downhills the highway path miles hurled under us. The cycling warmed me and I felt the bike respond easily to each pedal stroke.

Suddenly Antonia flew by me. "I am the *rayo veloz*!" She stuck her tongue out and zoomed ahead.

Behind her, Jake pulled alongside me. "Amazing! Just think: I carried a thorn from the slopes of Popocatpetl all the way to Oregon where it took a whole year to give me a *llanta ponchada*!"

"What an incredible tribute to your tire and the slime in your tube," I responded.

"I love slime!" he responded.

"Yes, we must not underestimate the virtues of slime!

Say, a minute ago I heard a madwoman from Mexico saying she was the *rayo veloz*. What's *rayo veloz* mean?"

"'Bolt of speed', I guess."

Slowing as Burnside turned uphill, Antonia slipped in between us. "Let's review. What did you learn today?"

"Nothing. Oh wait! I learned to respect slime."

She rolled her eyes and shook her head. "What else?"

"That you are the *rayo veloz* and you taught Lance Armstrong how to ride without training wheels."

"Fine," she sighed impatiently, almost ready to give up. We whirled through the Gateway waving at the MAX train. Pressed against the glass, three kids and their mother waved back. "Anything else?"

"Oh yeah. I learned that *llanta ponchada* means 'flat tire' in Spanish and *yabba babba babba* means 'Antonia found three pennies' in Jakeish. What did *you* learn today?"

"I learned that you are the worst Spanish student I ever met! And that you and Jake are children."

Antonia, Jake and I said good-bye before launching down the long series of Burnside downhills, where our paths home would separate. As we began to descend, coasting, standing on the pedals, Jake and I yelled question and answer:

"Yabba?"

"Babba babba!"

"Babba babba?"

"Yabba!"

I sped ahead, thrilling at the rush of thirty miles per hour down Burnside yelling back and forth to Jake, his answer coming over wider and wider distances—fainter and fainter: "Llanta ponchada! Llaaannnta Ponnchaaadaaa!! Llaaaantaaaaaaaa Ponnnnnchaaaaaaaaadaaaaa!!!!!"

traffic patterns

Stewart Clayten

zig then zag then pump on the gas
honk the horn and slam on the brakes
shake a fist then do it all again

single occupant vehicles going one way
along the my-way highway, snaking out
to the suburban sprawl, to big boxes of retailers
to strip malls with stall after stall after parking stall
lots of drive-ins, drive-thrus
conveniently driven to consume, to rush, to
sit in isolation, detached from
each other
"muscle" cars and SUVs idling in gridlock
each powered by hundreds of horsepower
—when just one will do

traffic flows
from A to B via
steel cocoons via
concrete and asphalt slabs
suffocating once fertile soil, stream and fields
an expensive toll
in a haze of combusting fossil fuels
layer upon layer

of soot and grit and grime
major arteries congested
arterial roads blocked
diseased society overcome by traffic
pumping emission, pumping noise
wheels spinning, pavement gleaming
slick with acid rain, with leaking oil

traffic patterns merge, collide
swerve around
the issues: green house gases, energy
depleted, junk heaps rusting alongside dusty roads
alongside road kill
discarded, ignored, disposable commodities

down a few liters, down a quart or two
down the block to purchase some more
short hops to the supermarket or corner store
conditioned to jump in a car
to drive two blocks and
back again

take small steps, left
then right then zig then zag
on foot

Empowerment

Rick Millikan

"Hello! Uncle here! I've got some business matters to be sorted out. Also I must take my car in for a tune up. It's running rough. Sorry, I'll have to miss brunch this week! Pass on my regrets." Such disappointing voice mail! Uncle told entertaining stories. So Anita would have to contend with Papa and brother Pablo without Uncle. They basked on the sun deck expounding life's meaning. Who would suspect survival might depend on breakfast banter and a bicycle?

Anita served fresh baked muffins and coffee. "Have a Java! Wake up! Smell the Lilacs."

"Our eyes are wide open, now let's reawaken our human potential!" declared Pablo.

"You'll perfect our lives? You philosopher kings need my practical input!" teased Anita.

"Certainly! Remember Anita when our family risked our lives to reform the 'Republica'! Here, we sit relaxed and well fed as we unravel life's predicaments! Who knows if our deliberations will reverberate and enrich this impoverished existence!" Papa smiled. "To think that Voltaire's 'Candide' inspired this garden!" Behold my green thumbs and look upon Eden. All seeds were planted lovingly, growing up with sunny attitudes, and now ready to overcome their stormy challenges. I am both gardener and transplanted weed. From a teenage hedonist, college utilitarian, and refugee stoic, I grew into a middle age pragmatist developing so many branches of materialism. Now my roots hold fast to spiritualism. Mindfulness evolves through fertile questions.

"So papa, what profound answer have you dug up?" prodded Pablo.

"Do not follow a good life. Lead a good life!" smiled Papa. "And to be empowered on the road of life, my answer is the bicycle!"

"Bicycle ..." laughed Pablo.

Anita joined too in the laughter. "Remember when we cycled to meet Papa's friend Dobby? Papa promised Dobby would give us the race of our lives! He did! We never cycled so fast! Dobby, the snarling Doberman, chased us and almost bit off our tires!"

"That was when I first thought about life and survival! Yet why a bicycle?" cajoled Pablo.

"My thoughts sing with bicycle rhythms!" spoke Papa.

"Doesn't a pensive walk get us equally where we want to go?" suggested Pablo.

"Walking may be your answer, but wheels of meditations carry me best to all destinations!" sang out Papa.

"So you're more than a cyclist?" queried Pablo.

"I cycle, therefore I am!" chuckled Papa. "Raise up your grape juice and toast: To bicycles, healthy, economic, effective green meditative machines, moving thought and persons successfully forward ..."

"Bicycles are fun, if you have time. I'm rushed! My car gets me quickly to where I want to go. It's also safer, more comfortable and yes, stylish." said Anita, lighting a cigarette.

Papa coughed. "Really?"

"Dear nicotine addict, no smoking please! An apple is healthier for all of us," insisted Pablo.

"Bossy Pablo! I'm not pregnant!" winked Anita.

"You know what I mean. Us! We breathe the same air!" Pablo tossed her an apple.

"I'll eat this juicy Spartan! But, I readily confess that these cigarettes get me through the day!" Anita stashed the apple and cigarettes in her purse.

"You smoke. Your car smokes. You want to poison yourself and your family!" Papa harshly frowned, then smiled again. "I do worry about you!"

"I know what I'm doing!" answered Anita stubbing her cigarette.

"Lovely daughter, sucking cigarettes may help manage stressful lives, but you know the unhealthy consequences! And that car presents another deadly habit!" Papa counseled.

"Even in my racy sedan, I'm going to be late!" joked Anita. "Thank you for giving your lovingly sage advice! Hasta luego familia!"

Papa hugged Anita. "Adios daughter! Drive carefully ..."

Her new cherry sedan soon sped away. Uncle's graduation gift of "wheels" had recognized her transition into a dynamic young lady. Uncle picked a car that reflected Anita's own qualities of reliability and elegance. Moreover, this sedan well represented her comfortable life and extensive freedoms.

Her critical family was very loving and generous. Papa and Uncle readily helped others and sacrificed everything for friends. When their comrades disappeared in the old country, they investigated government involvement. Being outspoken about their discoveries, they were branded revolutionaries and escaped to Canada. Here, Uncle ventured well, profited greatly, and spoiled Anita with gifts. Papa was also paid well for his creativity. Yet he anguished about his daughter's striving for "la dulce vida", the sweet life.

Heavy traffic made driving frustrating. "Freeway?" she sighed. "What is free about this way?" She slaved hard to support her crawling auto. What if it broke down? Could she afford its expensive repairs? If stolen, her life would be stolen. Stuck in traffic, life seemed jeopardized!

How does one survive such stress? She switched on the news. Item one detailed terrorism's effect on gas prices. It was bad enough to worry about being attacked. Now she had to budget for more car expenses. Item two noted a recent auto theft! Four teens were discovered in the midst of their joy ride. During the police pursuit their jolly stolen car crashed into two other vehicles. Imagine high speed car chases anywhere in this congested city! Item three related to air pollution's crop damage and consequential respiratory diseases. How many tons of toxins were being produced in this traffic snarl? The radio was turned off.

Cars growled and honked. All awaited signs of movement. No road rage dramas, just passive frustration. Checking her mirror, cars continued to line up. Anita lit a cigarette. She believed that cigarettes kept her thin and completed her cool tough image!

Her high marks in business courses had landed her a great job, soon she would be well off like her hardworking Uncle! Her Uncle was confident that Anita would be successful. She would soon be rich like her Uncle! Uncle Enrico started on his road to riches as a building contractor. After buying a car dealership, he joked about his new "exhaustive work". "Enough of this cold climate! We need cars producing a new, warmer world. As the water rises, I'll go back into real estate and develop beach front property!" He laughed and added. "Why worry? Be happy!" He was a hard working, hard drinking, fun loving Uncle! He loved fast cars, fast women and the fast life!

Her cellular phone zinged. "Hello Anita, it's Pablo. Sorry, I call you with bad news. Uncle ... Uncle is in Emergency. This morning Uncle smashed into another vehicle. He's on the critical list!"

Anita put out her cigarette, "I'll see Uncle after work." She couldn't take time off work! Uncle proclaimed never get behind in business. And, he also never gave up rich foods, quit smoking or found time to exercise.

"After work? That could be too late. Go to the hospital now! We need to be there!" Pablo demanded.

"Look, I'm stuck in traffic. But, I'll call my office and see you as soon as possible!" Anita reconsidered her loose ends at work. These needed to be tied up. Uncle would understand. She'd see him later.

How late would she arrive at work? Her mama's timely gift, an exquisite diamond studded watch, helped her make an estimate. Mama always had time for her, yet time ran out for Mama. Her mother avoided constant dangers before moving to Canada and her horrible auto accident. Someone hurrying to make a left hand turn, simply made a common, stupid mistake. Her mother's death became another precautionary motor vehicle fatality statistic. Did Uncle rush unthinkingly to his life's sorry end as well?

Cyclists passed by. She once pedaled off to school, to shop, to work ... Now she felt too anxious to wear unflattering spandex and deal with helmet hair. She spotted another lone cyclist on a tandem. Shaking her head in disbelief Anita said. "Papa, what are you doing here?"

"... rescuing you from your automotivated nightmare! Let's go see Uncle!"

"My work ..."

Papa continued, "Work's not life! Liberate yourself! Tell your boss about an emergency meeting with an important businessman! We need to pay our respects and give our best wishes to your Uncle!"

Anita hugged Papa. "My gallant, crazy Papa!" She parked her siren red auto off the shoulder. Then, Anita mounted the tandem's "stoker" saddle behind papa, her "captain". They smoothly pedalled off toward the hospital.

"Sing out!" encouraged Captain Papa. "As our wheels do roll along, roll along, roll along ... Let's sing out a cycling song, a ying yang song ... as we roll along ..."

Thinking was rejuvenated. Although Anita felt life's profound contradictions, she moved forward with a new clarity and muttered, "Is that car an asset or a liability? Could 'bicycle' be the answer?"

Abundance of Options

Karen I. Stanley

Accessory

Both of us unlock our bikes outside the gourmet grocer.
I say: we're alongside an absurd, giant vanity toy.
Extremely fine product of engineering, you say.

People charmed, stare slack-jawed. Metallic baby
blue pillow puffed up hard body
glowing like no other designer condom.

Well-suited driver emerges from shop
surveying the scene. Amused lips curl
keys tinkle signalling repossession of vehicular fantasy.

Bags balanced we ride
away in opposite directions
on the sidewalk (!)

Cosmetic

I must, I must
Enhance my
self image. Somehow
envision myself
accelerated. I'll
Showcase me. Encase me.
Windshield me. I'll feel free.

Have to get me
some recognition.
As in, I'll fit in.
Oh, I'll be the one. The one
with the perfumey, iridescent,
off-gassing, shiny painted can
compact-mirror machine.
Like me.

Finally I will be
grown up, or
I'll get propped up.
Of course,
With help from parents
I can be a star
Client

Maybe get a commercial

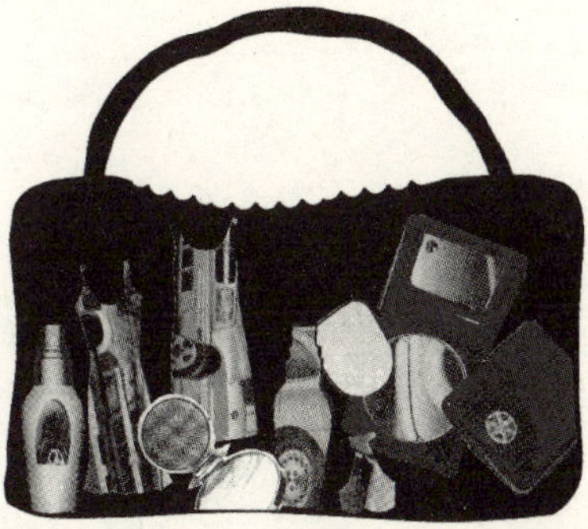

Remember

Outside is
Not in a car.
It is Blinking at Bright
Green Moss, Moving
Shored up by Wind,
Raining on face,
Breathing air
Is Outside

Too Busy Now

Transit used to work for people
Together.

Imagine
A time we spoke
Let body remember
how we got from place to place.

The War for the World: Three Sculptural Views

Jeff Mann

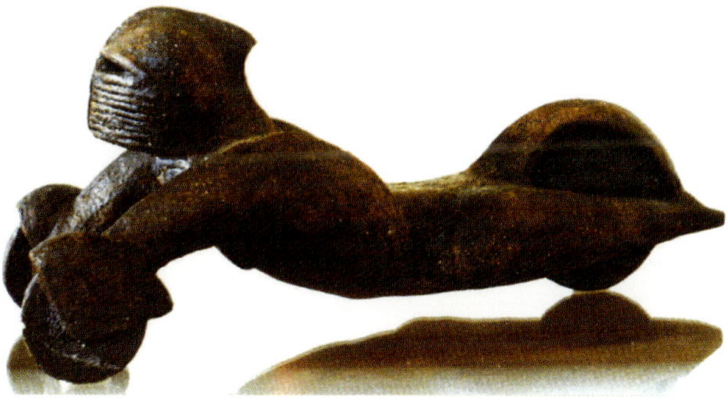

Automan, 1998, clay with stains, 16" x 6" x 5" (41cm x 15cm x 13cm). This was my first sculptural piece after getting out of production pottery. I think it's a rather bleak piece which expresses my feelings about how isolating cars are and how integral they have become to our lives.

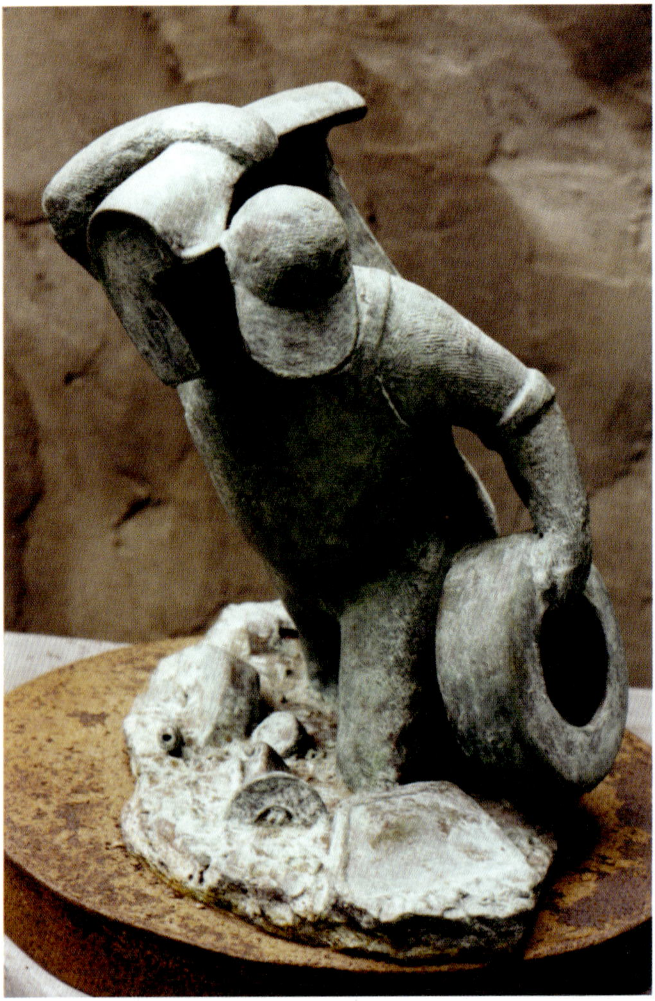

Parts, 1999, clay with metallic paints, 19" x 19" x 12" (48cm x 48cm x 30.5cm). This piece dictated its own creation in that I originally intended it to be three figures, but this figure came out so strong I stopped with him. I originally wanted to show the figures as being weighed down with car parts slogging through the mud in a kind of purgatory whereas this figure suggests a triumph over the machine which is OK with me.

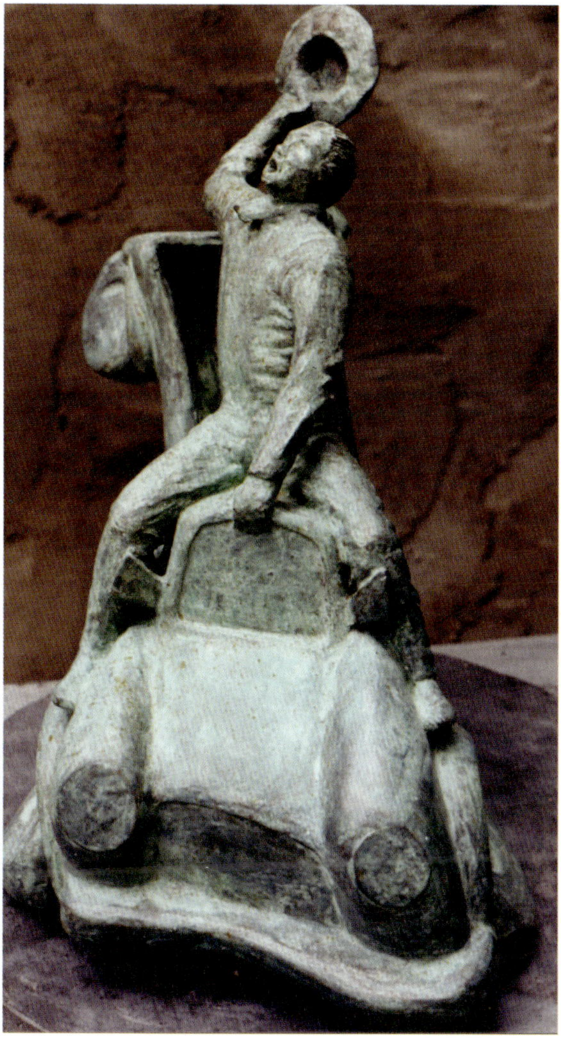

Remington, 2000, clay with metallic paints, 20" x 13" x 14" (51cm x 33cm x 35.5cm). This piece came to me as a kind of cartoon. Many truck ads are seeking to identify with the cowboy image so the take-off on Remington seemed obvious. The sad reality of course is that very few pickups ever really get used as the ads portray them. Most are used by commuters and sit in parking lots all day.

The Hunt and the Victory

Jim Hoehnle

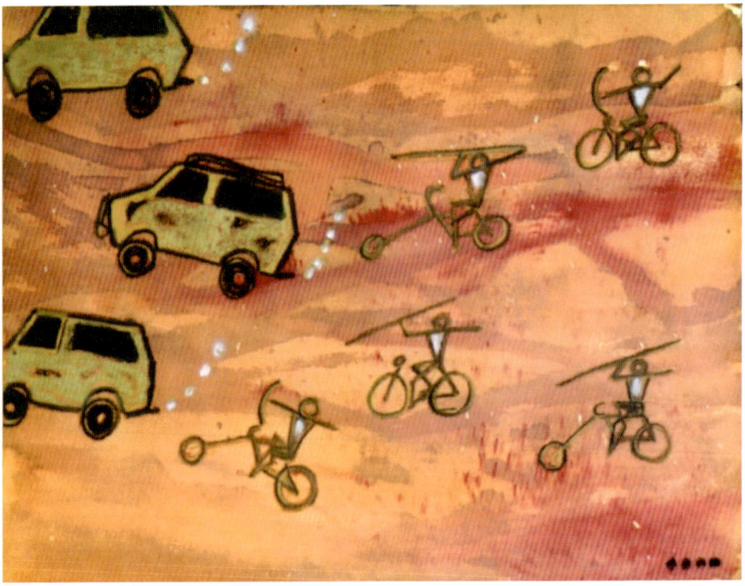

The Hunt, 2002, Mixed Media on Paper, 24"x30" (60cmx75cm).

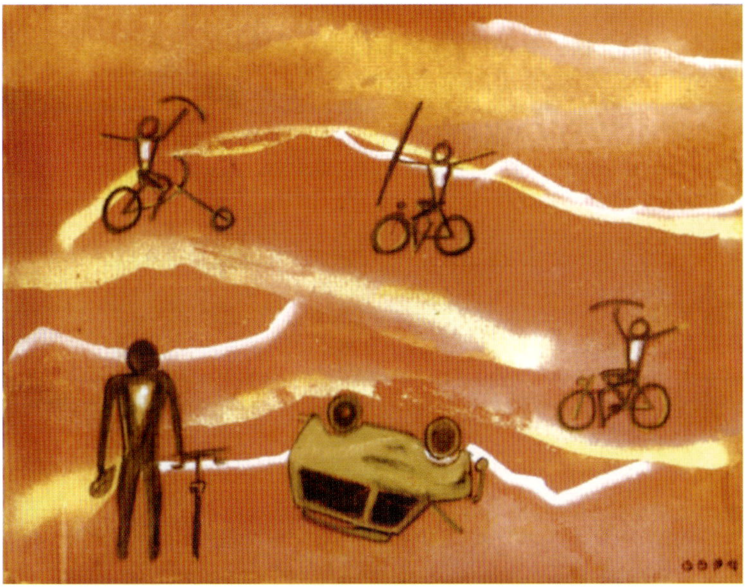

Victory, 2002, Mixed Media on Paper, 24"x30" (60cmx75cm).

It is an incredible arrogance that manufactures personal vehicles at today's scale and a tragedy, that the same arrogance is assured of their acceptance in the marketplace.

Driving a car is a personal decision and one of the consequences is an undeniable personal link to over-consumption and unsustainable living. Driving a car is not as inevitable as we are led to believe.

My hope in producing images like these is to help subvert this way of thinking and help bring about a time when the limited resources we have left can be put to best use.

Reports from La La Land

Sue Clancy

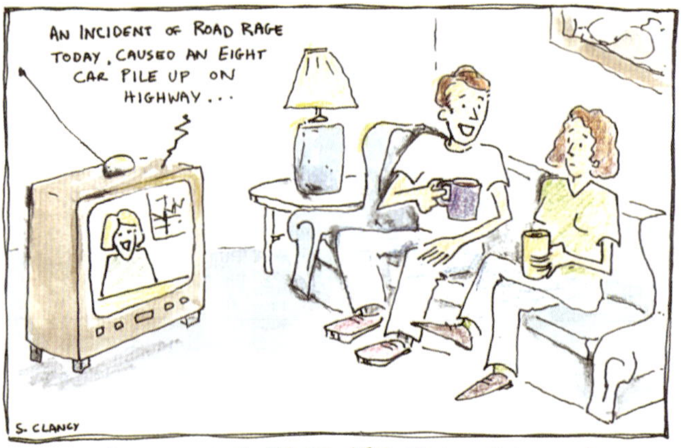

"EVER NOTICE THAT THERE ARE NEVER ANY MENTIONS OF BICYCLE RAGE OR PEDESTRIAN PILE-UP'S ?"

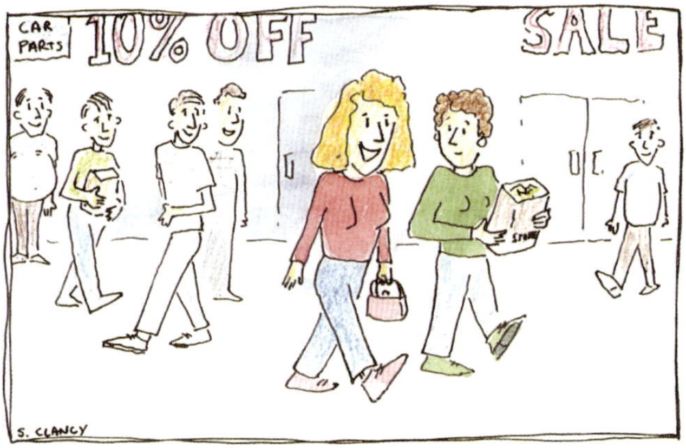

"I'VE FOUND A SURE WAY TO ATTRACT MEN. I SPRAY NEW CAR SCENT ON MYSELF! I GET DATES AND IT MASKS ANY ODOR OF REALITY!"

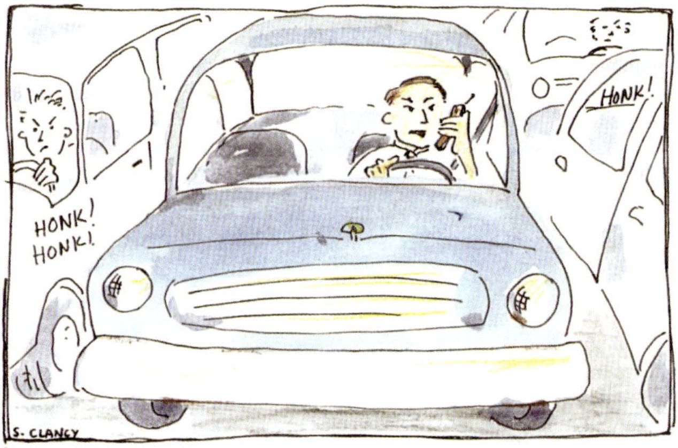

"I BOUGHT A HOUSE IN THE COUNTRY FOR SPACE AND QUIET, YOU KNOW? ONLY NOW I'M IN THE CAR IN HONKING GRIDLOCK TRAFFIC OVER HALF MY LIFE..."

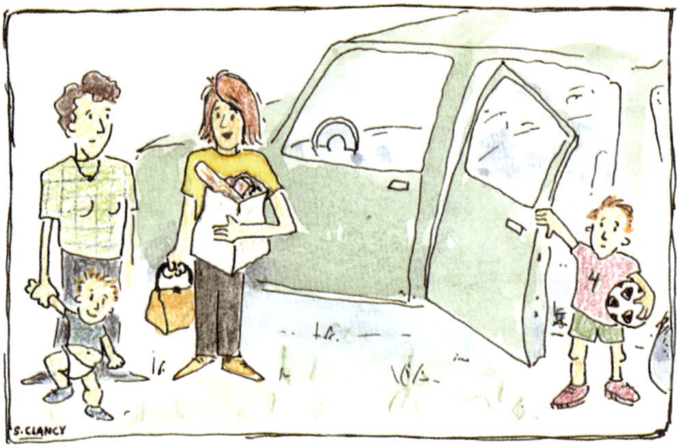

"HONESTLY BETH, I THOUGHT WE WERE MOVING INTO A HOUSE IN THE SUBURBS, NOT INTO A LUXURY S.U.V."

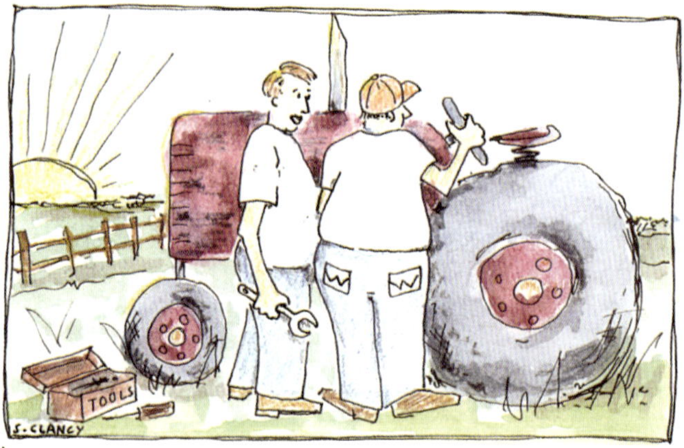

"YOU KNOW, IF IT WEREN'T FOR PEER PRESSURE I'D HAVE TURNED THIS GAS-GUZZLING CHRONIC BREAKDOWN OF A TRACTOR IN FOR SCRAP METAL AND GONE BACK TO WORKING WITH HORSES."

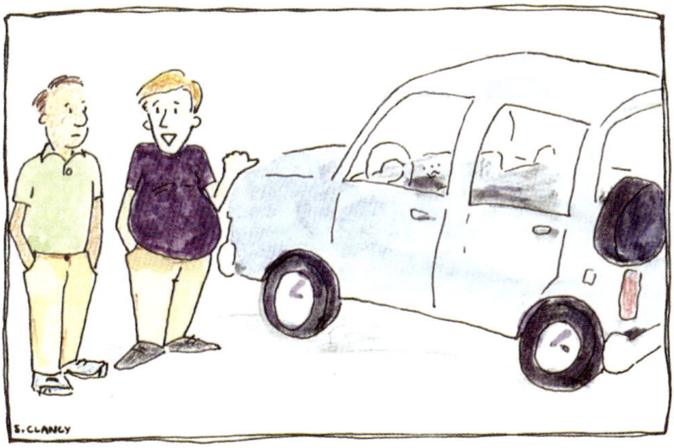

"I'VE BECOME DISILLUSIONED LATELY, MY NEW CAR HASN'T MADE ME SEXIER, WEALTHIER OR SMARTER"

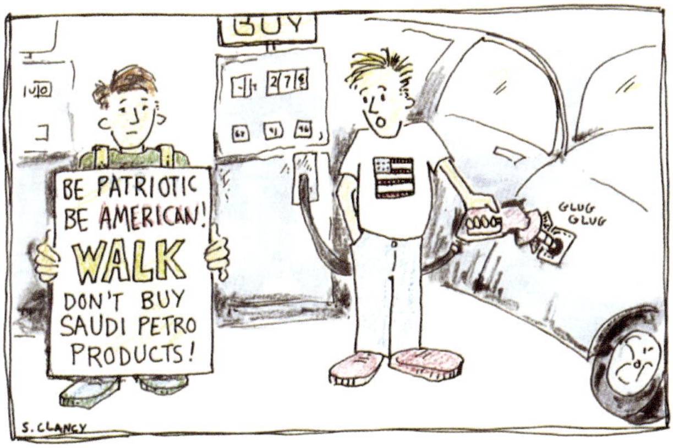

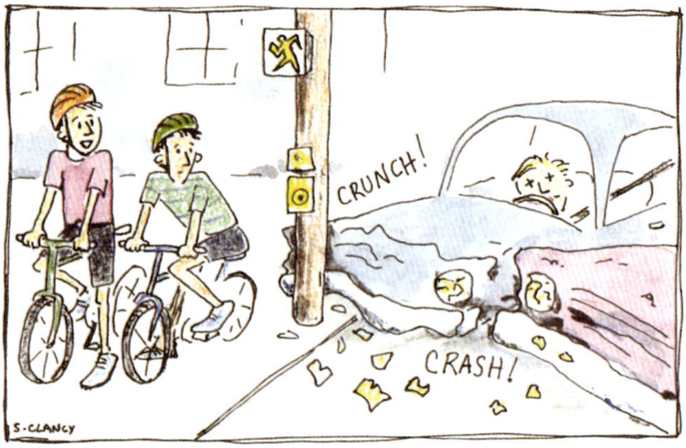

"SHOULDN'T THEY WEAR HELMETS?

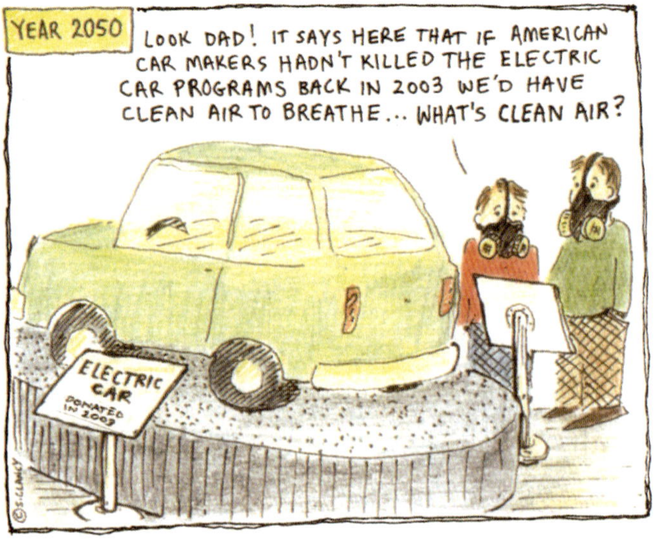

Point of Rest

RedSara

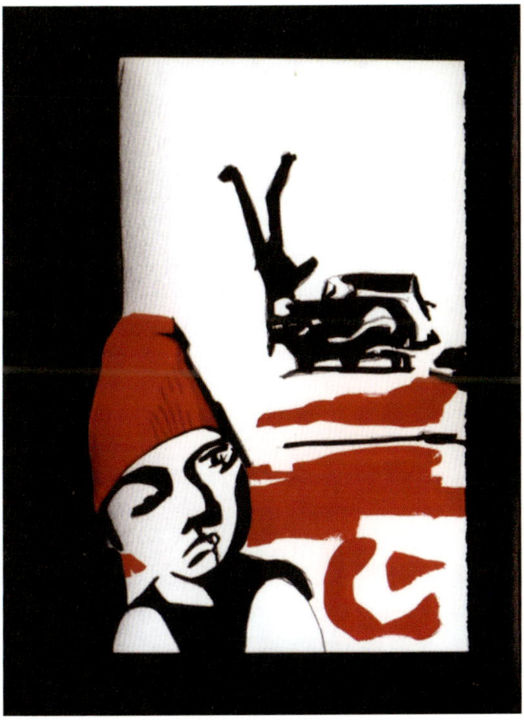

Witness, 2002, Mixed Media on Paper, 22"x30" (56cmx76cm).

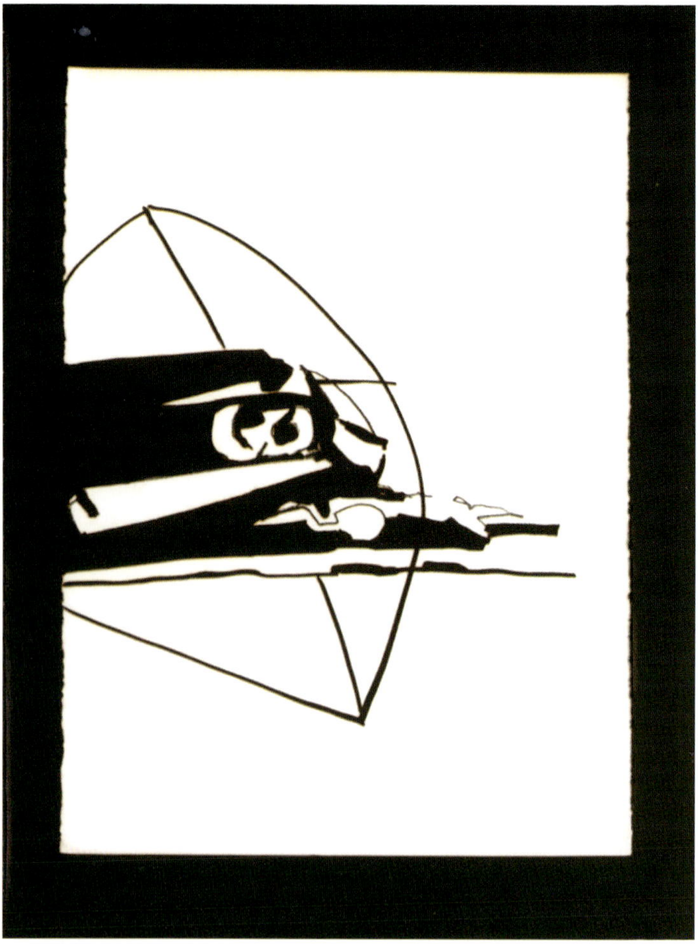

Impact, 2002, Mixed Media on Paper, 22"x30" (56cmx76cm).

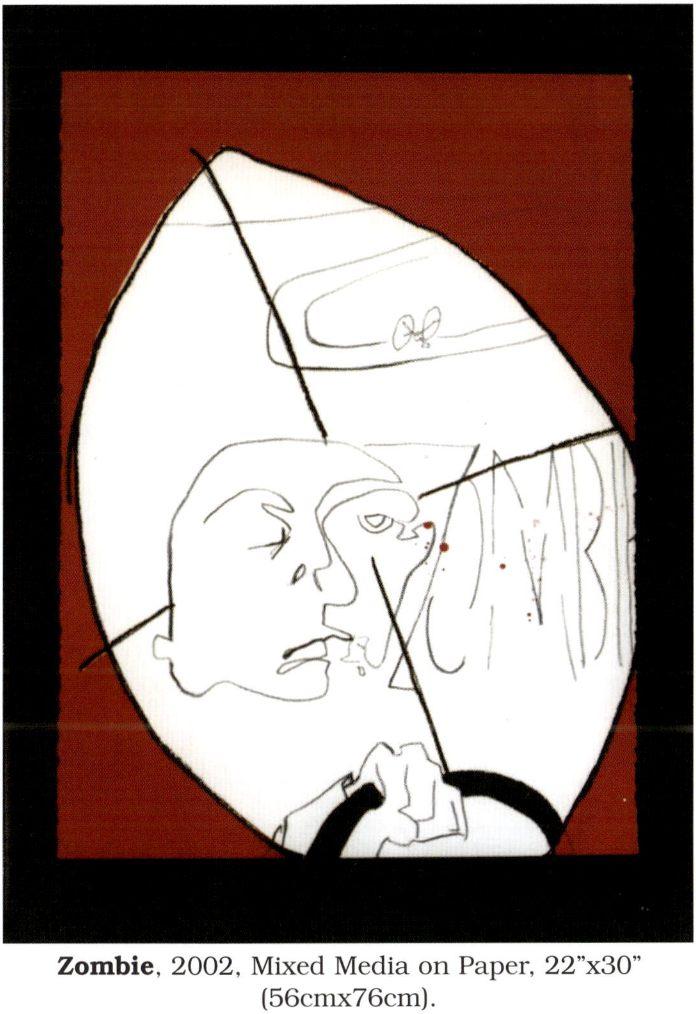

Zombie, 2002, Mixed Media on Paper, 22"x30"
(56cmx76cm).

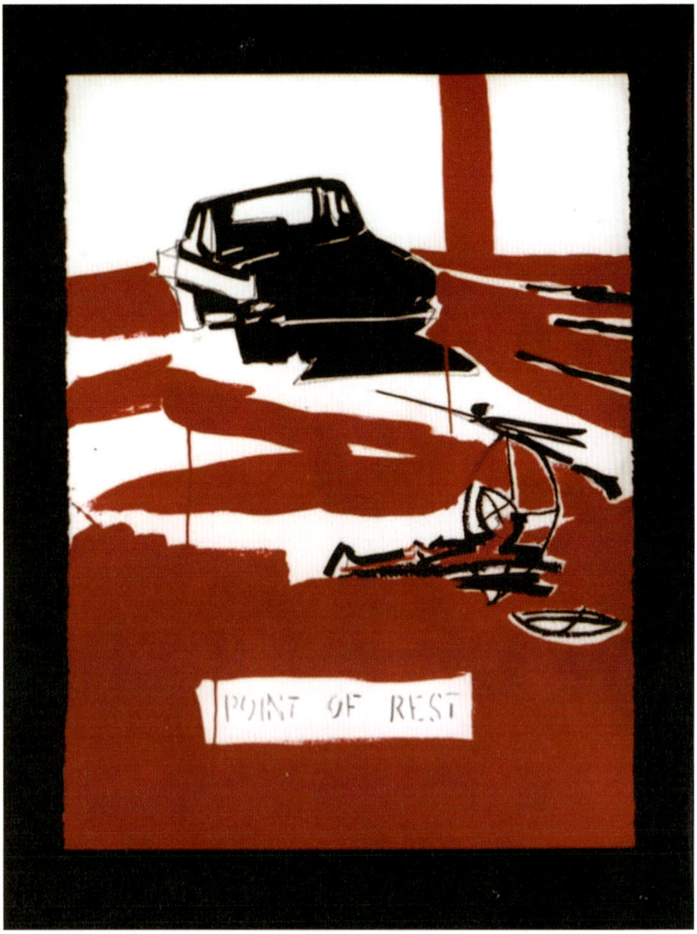

Point of Rest, 2002, Mixed Media on Paper, 22"x30" (56cmx76cm).

The Vector Project

Scott Massey

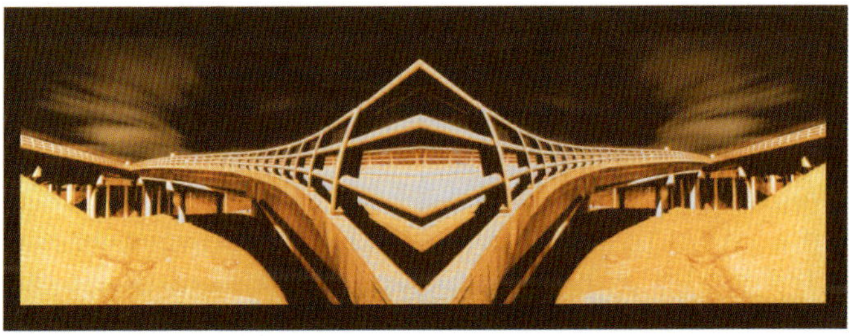

Vectors I, 2002, Aluminum-mounted metallic colour print, 16"x48"x1" (41cmx122cmx2.5cm).

The word *vector* is derived from the Latin root *vehere*, "to carry".

This series broadly investigates notions of movement; specifically constructions that purport to facilitate movement but are in fact part of a reality of increasing immobility. In this world an archaic system of transport continually doubles upon itself, rendering each subsequent layer less effective than the previous one, thus slowly undermining the entire system.

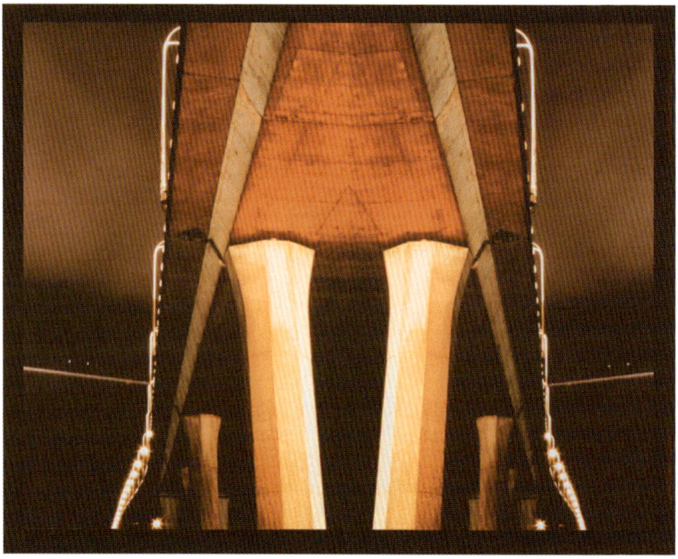

Vectors II, 2002, Aluminum-mounted metallic colour print, 24"x32"x1" (61cmx81cmx2.5cm).

By mirroring the elevated roadway structures back upon themselves, I am suggesting what is easily apparent from looking at a map: we have laid down more roadways than we know what to do with. All of these roadways were constructed for the express intention of facilitating the movement of vehicles and their occupant(s) from one place to another, point A to point B. Anyone who has attempted to travel from one point to another during high volume traffic is well aware of the failures of this system and must realize that constructing more roads is not the solution.

My Ride

Kenneth De Crescenzo

It is dark and I am cold. I shiver as I wait for the garage door to rise, and resolutely push my bike into the early morning air. An indulged, creature-comforted part of me begs to go back inside, warm myself with a hot shower, and eat breakfast before driving myself to work on four wheels instead of two. The more responsible part of me, bowing to the reality that I inhabit a world of lipids, cholesterol and high blood pressure, narrowly defeats the urge to resign the field before I have fired my shot.

As I straddle my bike and lift onto the seat, I adjust my fanny pack—my earthbound equivalent of a parachute, containing spare tube, tools, and clean underwear. I switch on my helmet-mounted light and its bright halogen beam slices through the mist. The only other sign of waking life is the neighbor's cat, whose eyes glow brightly from under a parked car, as they are caught by my light's beam. I turn on the flashing taillights on both my fanny pack and bike and feel the click of my shoe clips as I push off on my pedals.

The cool damp air cuts through my Lycra tights as I begin picking up speed. Once again, I must silence thoughts that would have me return to the warmth of the house, allegedly to don a pair of gators. Experience has assured me that within a few blocks I will have warmed up. That knowledge enables me to summon the motivation to push on. As I cross the bridge spanning the canal behind my house, I see the taillight of a car, the first sign of humanity I encounter today. Silently, I speculate about its driver, cushioned in the soft upholstery of the seats and embraced by automatic climate control warmth. I wonder if he or she has the radio on and if so, tuned to which station?

I pick up speed as I approach the first downhill, a low fall only two blocks from my house. It is a boost that I use to accelerate my spin and to set the pace I will try to hold until I am at work. At last I feel my stride coming on. Silver beads of mist build on the leading edges of my gloves. The tule fog hovers a few feet above the lawn of the local park as I glide past. Even now, less than a half mile from my house, I am starting to warm up. The temptation to return is now banished as I focus on gaining the bike trail another half mile ahead.

I swing out onto the main drag of La Riviera Drive. Occasionally a vehicle passes and leaves me in the wake of its taillights and exhaust. One whiff, and the exhaust odor wipes out the sweet cottonwood smell of the nearby river for the day. I pedal past the supermarket with its delivery trucks parked alongside, and turn right onto the driveway that leads to the American River Bike Trail. There is now a little more daylight than when I set out, and in the early grayness of dawn I can make out the tree line by the river.

I pedal vigorously as I reach the downhill path that connects with the trail. This is my first personal best of the day: can I beat my own record? I focus intently on my speedometer as I speed down the hill. I grasp the brake levers in case I encounter someone or something at the bottom of the slight curve below. The numbers rise swiftly and by the time I reach the bottom I am moving at double the local speed limit. Fast, but no personal best. My challenge now is to hold that speed through the first rise and use the momentum to carry me through the straightaway that lies ahead.

Out of the saddle now, I pump through the rise. At the top I encounter for the first time today, another person not encapsulated in a machine, a jogger. He waves and I shout a morning greeting as I zoom by. In our cars we don't make that sort of personal contact. We become machine-like ourselves, passing each other's vehicles with no intentional cognizance of the human lives they transport. Somehow, the chrome and the leather distract us, and a fleeting chance to smile at another person, to say good morning, thank you, or a simple hello is lost to darkly tinted glass, brightly painted steel and the throaty sound of a rumbling engine. These thoughts sometimes sting my philosophical side during my morning bicycle trips.

I pedal farther down the straightaway and beneath the power line crossing. I now enter my favorite part of the trail, the Enchanted Forest. It begins with a slight descending "S"-shaped curve between rows of wild grape and blackberries. I gain speed at the top of the slope. At the bottom, a bit of mist hangs over the trail as it curves, this time rising to the left. I stand again to tackle the rise. At the same time I peer ahead for wind-blown branches or leaves that could pose obstacles or slippery patches. Seeing none, I continue to pump hard, breaking a real sweat now as I labor through the tree-enclosed trail.

Emerging from the woodland, I notice cars backing up on the Howe Avenue Bridge. I smile to myself as I contemplate what their occupants have missed in their day. I also secretly rejoice in that, for if they were on the bike path with me, my trip would be harder and perhaps less enjoyable. I pass over the access road and under the bridge. I look for the homeless man who lives at the junction of the bridge and its earthen support. He lives in a hole surrounded by rocks to hide his presence. The handlebar of his bicycle shows above the rocks.

It is much lighter now, and I can see the beginnings of a golden glow on the treetops as I pass the water treatment plant and enter Alumni Grove by the college. The orange paint of the Guy West Pedestrian and Bicycle Bridge takes on a deeper tone in the dawning light of the rising sun. In a single switchback, the trail climbs to the top of the levee and now my view encompasses the buildings of California State University Sacramento and the Fair Oaks Bridge about a half-mile away. The river is a perfect mirror to the sky, reflecting its early morning blues, golds and grays as well as the deep greens of the woodlands that line it. A pair of geese skims the surface as they land, leaving the v-shaped marks on the water, and I wonder if any of the people in cars on the bridge have had a chance to witness their landing, to appreciate this charming sliver of nature in our city.

I prepare for my second challenge of the ride. Passing through the bollards on the Guy West Bridge I step up my pace. I am working for my personal best in crossing this level stretch of levee. Once again, the numbers on the speedometer start to rise; however, this time they are not aided by gravity. I hear and feel my breath laboring as I pour as much energy as I can into each stroke of the pedal. The pavement speeds beneath my wheels and each campus building I pass becomes but a blur in the landscape. The bollards at the end of the levee approach and with one last glance at the speedometer, I apply my brakes to prepare for the sharp curve that will take me down off the levee and into traffic.

I skim along the edge of busy J Street in front of the college, crossing under the railroad overpass. I approach the first really dangerous part of the trip: a blind curve off ramp that I must cross. Unclipping my pedals for safety, I crane my neck to see as far as possible before dashing across the paved gulf. Even though it is a cross walk, there are no signs to warn traffic of my presence; an ill-timed move or an inattentive driver and I become another bloody statistic.

The reserved bike path along M Street offers the last refuge from traffic on the trip. For the next two miles, my route is on a residential street that is parallel to two main thoroughfares. Even though traffic is light, I have to be careful with the cross streets as cars run stop signs often enough. I must admit that I often am not too careful with them either. Still, this route offers a last chance to pass through an environment where once more I can feel civilized as I continue my commute. I ride by as semi-sleepy school children and workers leave their well-kept homes to start the day. Occasionally another cyclist will join me and either we will pace each other or engage in a casual morning conversation as we ride together through the waking neighborhood.

I turn onto Folsom Boulevard, merging into heavy traffic without the benefit of a bicycle lane. Now I must share the road with cars, trucks, buses and their drivers who may be less than half aware of my presence. As I merge on their right side, not immediately in their line of sight, I fear for my safety. My life now depends upon the flashing lights on my back, my vigilance, and the brilliant neon yellow of my jersey. The place where I join with motorized traffic is particularly treacherous because two lanes merge into one. I have seen many a drag race as two side-by-side motorists jockeyed to get ahead of each other, and I have wondered if they thought they were competing for the pole position at the Indianapolis 500. I pick an open spot in the traffic and pedal hard, hugging the edge of the parked cars, hoping that no inattentive driver will throw open a door into my intended path.

At the 30th Street traffic light I pull up, rest my leg on the curb and look over the occupants of the SUVs, luxury coupes, and lesser cars that are also waiting. Through the safety glass and sheet metal we regard each other. To their drivers and passengers I am outlandish, wearing a bright neon yellow jersey, pearl white helmet with spotlight mounted on it, black Lycra tights, and adorned with numerous flashing red lights. To me, they appear equally odd with their hairspray brittle coiffures, makeup, dark suits and power ties cut off from the freshness of the morning. It is so different from the joggers and cyclists on the earlier segments of the trip. No hellos, no pleasant conversations. On a good day, if I am lucky, I can elicit a nod or a brief smile.

Riding in traffic is not an easy thing for me. I know what it is like to be hit by a car. Fortunately, my car/bike encounter caused no personal injury although it did total a ten-day-old bike. Again, fortunately, the driver's insurance covered my costs for replacement. I often think about the humor of the incident. I was hit by a State Senator as she was exiting from the Capitol Garage. I was in a bike path and had the right of way. After I flew over the hood of her car, I landed on my chest and stomach. As I lay face down on the ground, letting my body parts check in to let me know if they were unhurt, the Senator came running out of her car yelling, "Are you hurt, are you hurt, I am a Senator." I often wonder what she thought her station in life would have mattered if I had been more seriously injured. If anything, the incident has made me more aware that rich or poor, famous or unknown, highbrow or not, a moment of inattention can have disastrous consequences for someone whose only protection from the pavement is the fabric clinging to his or her skin.

I arrive at work and chain my bike to a rack in a secure cage. Walking to my office building I feel like some astronaut venturing back from a flight in space, helmet tucked under my arm connected to my pack by a single wire for the headlamp. People who know me have become used to the sight. Others often take a second glance. I mingle with my coworkers some of whom are getting off the light rail train at the station outside our main entrance. Others are coming from the garage. They head to their offices and I make my way to the showers and my locker. As I dress for the day, I share my experiences of the ride with my cyclist coworkers. We comment on the briskness of the morning, the brightness of colors of the sunrise, whether or not we saw anyone, and sometimes we share any near misses we may have had with anyone with whom we share the road.

Roadkill

Jeff Younger

When setting out to write a thematically appropriate jazz tune for *Traffic Life*, a few basic characteristics of modern day traffic immediately came to mind. Parallelism, angularity, perspective and stasis/motion found their way into the form and both melodic and harmonic content, proving to be an interesting and fruitful set of concepts with which to begin. The visceral experience of being caught in traffic, a more detached bird's eye view, and wisps of philosophical musings on rhythm and technological development round out the more abstract inspiration and intent within the music. In performance, Roadkill will swing, fast and relaxed, revealing an odd beauty in its angularity and purposeful peculiarity.

The form is AABA, and as though looking down a stretch of highway toward the horizon, each section is shorter in duration than the one previous. The first two A sections are 32 bars in total, the B section half of that, 16 bars, and the final A section 8 bars, half the length of the B section. The harmony in the first two A sections is fairly static, with parallel structures moving every four bars. In the B section, the harmonic rhythm and contrast increases, but seems to stall over the section's final 8 measures. For the final A section, we return to the first chord of the piece and the

(continued on page 124)

ROADKILL

JEFF YOUNGER

UP SWING

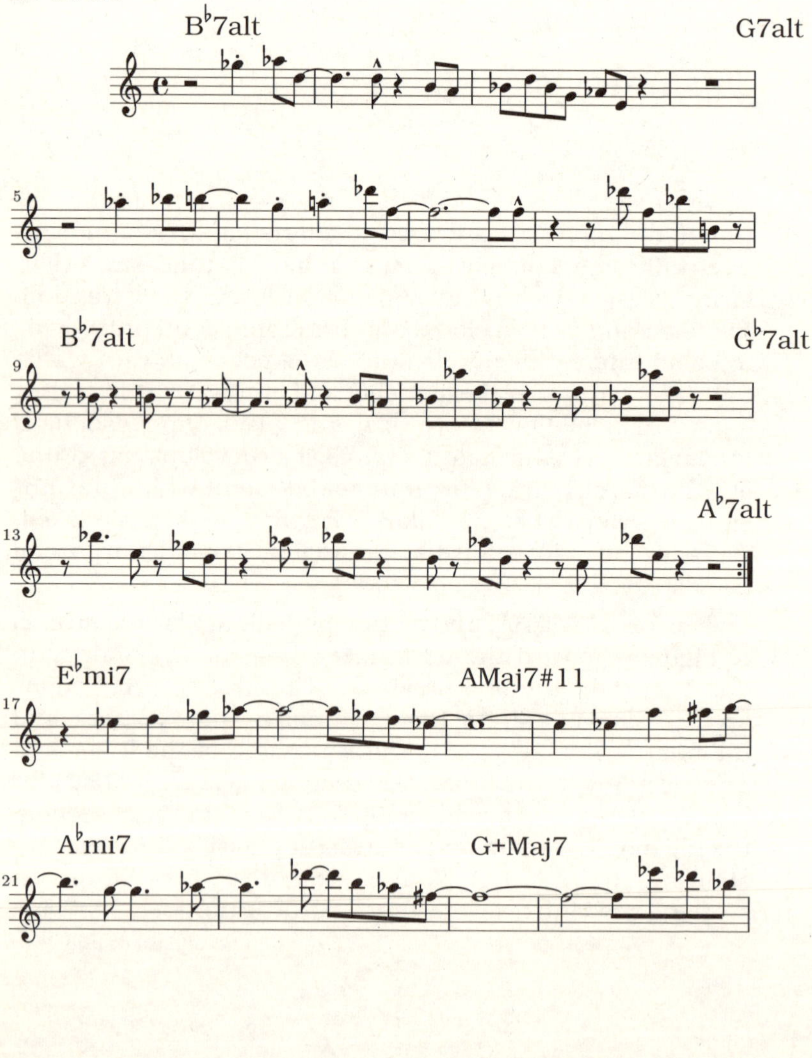

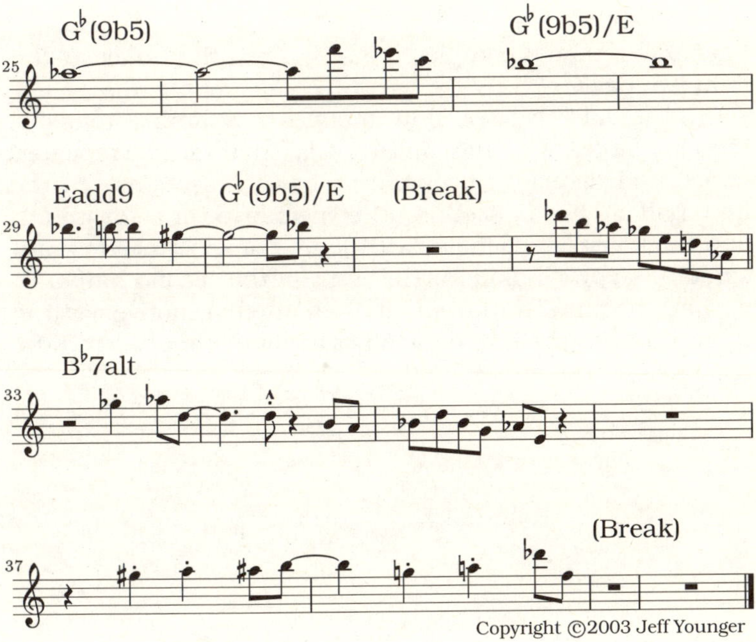

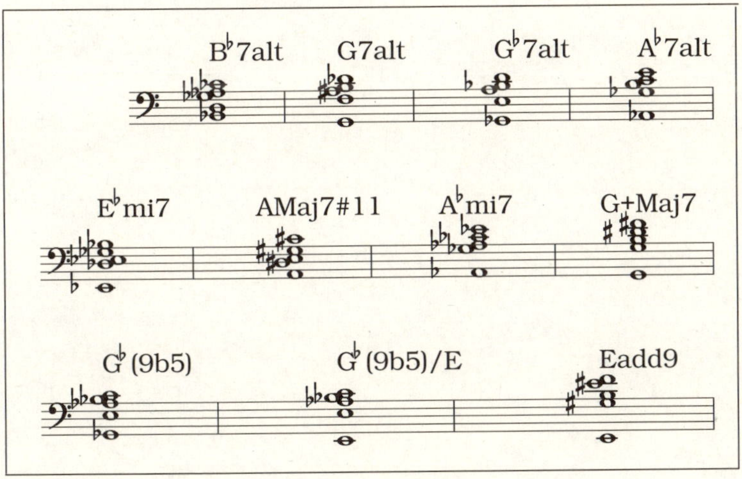

Figure 1. Example chord voicings.

(Continued from page 121)
harmony remains completely static until the break at the final two bars.

In the melody, we find motives containing tension-inducing notes and angular intervals, rhythmically repeated, varied and displaced over the first two A sections. In the first half of the B section, the melody begins to stretch and move more flowingly, with long, consonant notes and smooth, stepwise motion. However, in the second half of B, we find the parallel movement of an angular motive, setting up a return to the initial melody during the final A, and the conclusion of the tune.

First, you buy a car

Robert Gregory

First, you find the car of your dreams at the local motor vehicle lot, where else? You know that you have been without a car for far too long. You can't just keep on riding with friends, and the public transportation system has long since gone out of business.

Here's one, a lovely motor vehicle, replete with bells and whistles, with pretty colours too. Grasp the door handle, pull and open the front door. Then sit in the front seat, and note, the seat feels just right. This automobile smells so clean and new, the dials and gauges and wheel and seats look pristine. You decide to buy this vehicle.

Then you hop out to talk with the salesperson. The sticker price seems about right, so you offer to pay that much, of course though, via so many payments over so many months (actually it's years but they always list the months). The salesperson smiles, so you know you have made a good purchase that will serve you well. Isn't a happy salesperson always a good clue to an efficient and reputable deal?

After all, the advertisements in the newspaper and over the television always show how much fun people have in their motor vehicles. You have to fill out the papers, but the salesperson makes that part easy. The salesperson tells you just where to sign.

Second, you sit back down in the motor vehicle, turn the key, start up, and drive away. The advertisements promised so much, and you quickly find yourself out on the highway, cruising along and as happy as a lark. You turn on the radio and sing along, not noticing the sign about speed cameras operating in the area. Eventually you drive home, and look forward to many more trips from here to there and elsewhere.

Once you have bought the motor vehicle, other things seem to flow right along too.

Thus, you have to buy petrol. The petrol has to be of the right octane level, but that is easy. The people at the petrol station know. They are happy to help. You pay with a smile. You still wonder a bit though about paying Goods and Services Tax of 12.5 percent, on top of the cost of the petrol and the government tax. That works out to be a tax on tax, which should be illegal!

And oil. The oil has to be of the right viscosity, and even though you don't know quite what viscosity means, the people at the petrol station do. They must! And you have to remember to check and make sure you have water for the windshield and for the radiator. You pay their bill for oil too.

Then you suddenly think about your license to drive. It has been a few years since you got a license, and you have to be sure your driver's license is up to date. No problem, you just know you had one a few years ago. It must be around somewhere.

You notice too, that you need a new motor vehicle license plate, for the one from the motor vehicle lot belongs to the lot and the salesperson wants it back tomorrow. You have to find the location of the motor vehicle registration place, where they happily tell you that you need the title and your driver's license.

And, of course you have to get the insurance for the motor vehicle, and for yourself in case you run into someone or perhaps if someone runs into you. And you will need insurance for uninsured drivers for if they don't have insurance, then you will still be covered. You go to the insurance agency, and they explain patiently that you have to show your driver's license first. The insurance sales people all have smiles on their faces. You pay their bill, knowing that all will come out favourably. After all, you know, "she'll be right!"

Thus you go home and search and eventually find your license but note that it expired a year or so ago. You ring the transport people and get scheduled for a driver's test for the new license, but you have to wait three weeks. They are quite busy. "Bring your mate along with you," they say. Naturally you are disappointed, but it is only three weeks and you will be able to drive your motor vehicle again.

In the meantime, once you have the motor vehicle, then you need a garage. You look for a place to rent near your home, as the street does not have sufficient space, and after all, you would not want someone to steal your motor vehicle. When you find a place, then you move the motor vehicle there. The owner smiles. You pay the bill.

Of course, you also get a lock to place across the steering wheel when you park somewhere, so no one can steal your beautiful motor vehicle. The automobile accessories shop personnel smile. Again you pay the bill.

Then you need waxes and polish and cleaning equipment. After all, a motor vehicle should be kept looking lovely, and dirt does damage the finish. Again, the automobile accessories shop folks smile at you. They seem as happy as the proverbial "box of birds!" You pay their bill.

You spend a couple of days polishing, rubbing, and coaxing the best finish out of your motor vehicle with new creams and waxes and shines. You pay extra for the best quality finishes, as you would not want just soap and water. A good finish is "as good as gold!"

Then you figure you might need an inspection by the mechanics at the local automobile association, just to be sure. They note that the tires are bald, and must be replaced, otherwise you can't get a warrant of fitness and a registration. You pay their bill for inspection, and watch them smile.

Then you go to the tire store and buy five new tires, one for a spare. They will install them for just a bit more, and of course, balance them too. The salesperson smiles. You pay their bill.

At least the automobile association provides you with maps of interesting places to go, and a card showing that you are a member. You pay that bill too. And, they smile. The map points out how New Zealand is "clean and green!"

When you go for the driver's test, you bring someone who does have a license, so they can't accuse you of driving to the site of the test without a license. You take the written test, but fail the vision test. Your mate is disappointed. So you have to go to the doctor's office and take another vision test there. The doctor charges $100.00 for that test, and says that you need glasses. Of course, this new discovery means that the glasses are really necessary, so you select glasses that are marked $400.00. You pay your mate something for helping, you pay the doctor's office, and pay for the glasses. Everyone smiles that day.

A bill arrives from the police claiming that you were driving at a rate of speed much greater than allowed. This refreshes your memory about your cruise when you first tried out your motor vehicle, and you recall that you might indeed have passed a traffic camera and you might indeed have been clocked at what they claim. You pay the bill. You write a letter to your local parliamentarian about the quality of the roads and the issues raised by traffic cameras. You can be sure they smile at your letter.

You can then return to the driver's test locale, where they schedule you for the next part of the test in three more weeks. You use the time to polish the motor vehicle again, and find that the windscreen wipers need replacing. You go to the automobile accessories shop and they smile when you enter and laugh as you exit. Of course, you pay that bill.

After three weeks, you go with someone just in case, to take the test. You drive around town and into rural roads, and back up and park and do all the other things that one must do to get a license. The assessor says you will hear in the mail about whether you passed or failed. The assessor smiles. You pay the bill for your mate and the bill for the assessor.

In two weeks this time, you receive a message in the mail that you failed, and are automatically scheduled to take the test again, in three weeks. The bill is actually a receipt for the first test. You pay your bill, just in case.

In three weeks, you go for the test again. This time you drive around town and then to rural roads where you back up and park and do all the other things that one must do to get a license. The assessor says you will hear in the mail about whether you passed or failed. The assessor smiles, so you feel positive. Hopefully, you passed. And you pay the bill.

In two weeks, this time, you receive a message in the mail that you passed, and they enclose a bill. You pay the bill again, just in case.

Now when you go out to drive, you remember the speed camera and avoid it. However, the police also moved the camera to a new, hidden location, so you again, in two weeks time, get another ticket for speeding. No one told you but you now have two tickets, and the three tickets and out system is a sad reality. You pay the bill.

Now it has been about 4 months since you purchased your vehicle. You are finally ready to go for a ride, across country. It is time to get into the traffic. This is what you have been waiting for. You smile at the anticipation of it all. You are happy to be in this lucky country, where life is carefree and happy, like the advertisements.

First though, you fill up with petrol. The station attendant smiles as you pay your bill. You still wonder about the taxes on the tax.

On the road! Finally, you drive away, perhaps from Wellington to the north, or from Auckland south. In Auckland, you clear “Spaghetti Junction” and enter the Southern Motorway. On an average day, the traffic stops for a couple of hours while overturned trucks are towed away, whereas going north from Wellington, the traffic stops for a couple of hours daily while head-on wrecks are towed away. Helicopters keep busy in both locations, ferrying passengers to emergency rooms in hospitals, while the Accident

Compensation Corporation accountants keep busy figuring out the costs, expenses, overheads, and most important of all, the personal super-annuation benefits they will receive.

New Zealand roads are a bit different from those in the rest of the world. The accident rate is about double, sometimes triple, anyplace else in the world, for the roads are narrow, poorly designed, terribly marked, and woefully kept.

But you finally get on the road, and fail to notice the little sign that a speed camera is lurking nearby. Past that, the inevitable happens. You get a ding, or a prang, when a vehicle in front or to the side stops too suddenly. Your bonnet and the other vehicle's boot are damaged. No injuries, thank goodness, but your vehicle is a bit of a mess. You are thankful that you have insurance. The traffic officer helps sort things out, and the tow truck pulls your vehicle away to a garage. The panel beaters look and study the situation, knowing that parts cost a lot. They smile, happily, as they anticipate their next week's wages. You pay their bills.

Finally, a week or two later, you get back on the road again, and sail down the highway. You are busy tuning in the radio, listening to kiwi-music! This time, there is no sign, but a traffic officer notes you travelling along, just a bit above the speed limit. Calling up your past history on the police computer, you are quickly notified that you have now four speeding offences in six months. You are promptly given a summons to court. You anticipate so you hire a solicitor (lawyer), and pay that bill, as requested, before the services are delivered.

On your drive to the court, you again run into someone who stops short. This time you have a more serious prang. In fact, the vehicle is totalled. At least you have an excuse for arriving late to the court. You pay the bill for the tow truck, the extra not covered by your insurance company and an extra $50 for the annual police benefit ball.

While you are waiting for the next date for the revision of the court summons, you study your situation. You notice, even though you have a warrant of fitness, a driver's license, insurance, and other necessities, you seem to have forgotten, in all your busy-ness, to follow up and get a registration. You have been driving an unregistered car! They will exact a price for that for sure. It looks like a jail term!

Then you add up your expenses thus far. Your motor vehicle has cost you more, much more, than you make at work, and you decide that you have to declare bankruptcy. You are now, more than ever, certain that with an unregistered vehicle, you will go to jail. But on the other hand, you know that even though your vehicle is now totalled, you have made so many people happy, you have put so much enthusiasm and money into the venture, that just as soon as possible, you will certainly go out and purchase another motor vehicle.

After all, you could just as easily have bought a horse, or a plane, or a bicycle, or a place in the local mental health asylum. Once you become part of any system, you are indeed part of the system and subject to all of its requirements and responsibilities.

Life In Transito, Ergo Sum

Rick Millikan

Though I'd rather be cycling to work or better yet in the Loire Valley, the Vancouver Skytrain offers another welcome option to city grid lock. Plus, a scenic perspective is always provided. Riding rhythmic rails can inspire journals of journeys and tales. I ride, ergo I think. Lives exist in transit. Today my laptop feels the keystrokes of a heartfelt tirade to rally rage against an auto age. Odious odes are unloaded to erode and explode those auto motives. Hopefully, such a moving opus inspires hope for us.

Skytrain Poetry
An Aerie View of Smog Demons
Beneath us scuttle metallic farting bugs,
That blur the eyes and tweak the nose.
These puffing putters roll rambunctiously in ass fault rows.
Enraged or just engaged in 'so what' shrugs,
These drivers strain with sour coffeed mugs.
And drive and dive dangerously into denty woes,

We ride supreme, above those hooting hordes,
We human poets enter Metro, ready to add reason to rhyme,
Whilst carsters creep, we are moved by urbane design!

It's not clear where my fellow passengers are going, metaphysically or functionally. Before I get off at my stop to see my honey, I speculate. Could fellow passengers be ardent shoppers prepared for mall mayhem? In the spirit of this green express, I hope not. It's more likely all of us are workers that may build a better tomorrow. Our first step today was boarding this train. We all could be students studying human progress, reading the writing on the wall, and empowering a green consciousness with our own life journeys. So what are we?

Sky Trainees, Life Cyclers
Whatever we are, we're not in a car,
Stylishly infatuated, encapsulated, isolated ...
Where we are, carries us far ..
Moving together in time, space and mindfulness ...

Boiling Point

Joanna Emery

With an illustration by Kevin Cook

Why did we stop?

I don't know.

How come the car isn't moving?

I told your brother, I don't know.

Hey, what happened? Are we there yet?

No, we are not.

What happened?

I said, I do not know. Sssh kids, your father has to concentrate on his driving.

But Mom, how can he be driving if the car isn't moving?

Look! I'm in the driver's seat so I'm driving.

I think you're just wasting gas, Dad.

Yeah, maybe we should have taken the other route.

Now kids, don't gang up on your father. All the roads are busier in the summer.

I don't want to spend all day in this red-hot car! I'll be fried to a crisp!

Me too.

Me three.

Fine by me! Next time we'll stay home!

George, watch your blood pressure. You don't have to be so harsh. They're just children

So what if I have a heart attack. They're driving me crazy anyway!

That's funny, Dad. Har-dee-ha-ha.

Enough!

But Dad, it's really hot in here ...

It's not my fault we don't have air conditioning!

We do at home, Dad.

Yeah, being stuck in traffic with no air and a zillion degrees outside sucks bigtime!

I wanna go home!

We can't go home, sweetheart.

Why can't we just turn around?

That's illegal.

I think it's illegal to leave your kids in a boiling hot car.

We have NOT left our kids in a boiling hot car! Your mother and I are here too!

Okay, so you feel as hot as us?

Of course! And since I'm bigger—I feel hotter!

So if we're going to die of heat exhaustion, you're the first to go.

No, dummy, the youngest ones go first. And you'll be second!

Hey, if you keep talking to me like that, I'm the Dad and I'm the biggest—I can make you go first!

Kids! Please be quiet! Remember your father.

Mom, Dad, none of this matters IF WE ALL BOIL TO DEATH IN THIS CAR!

Mommmmmy, are we going to boil?

No, dear.

But I'm so hot.

You're mother is right, sweetie. We're not going to boil but I HAVE HAD IT with this driving. I'm pulling this car off the road.

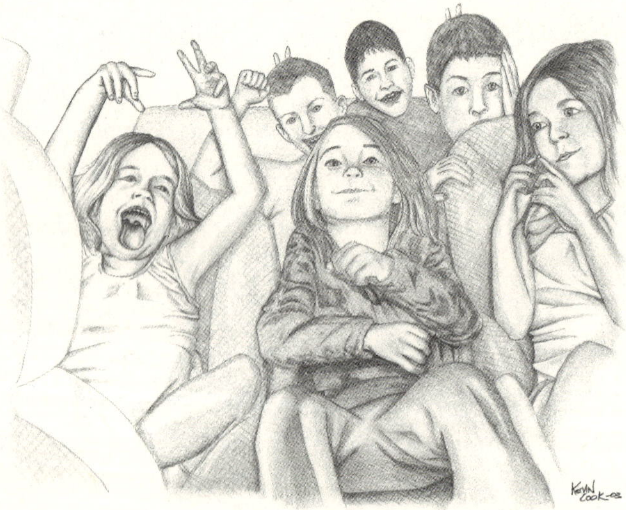

ALL RIGHT!

YAHOOO!

Still hot outside, but at least we're not stuck on the pavement.

And we can get some shade over by that tree.

Ah, smell the fresh air. No more stinky truck exhaust blowing in my window.

I'm just glad to be alive.

That's nice kids. Maybe next time we'll take the train.

Yeah, it's got air conditioning.

And no excuse if Dad's grumpy—he won't be driving!

Parking Structure Three

Wes Alderson

Something was wrong with the parking structure. It hurt John's eyes when he tried to look directly at it. At first glance, the gray concrete building seemed perfectly ordinary—a typical square, four level parking structure just outside the Psychology building of U.C.L.A.

But on closer examination, it resembled some of the illustrations featured in the stupid psychology book his wife Millie showed him. While Millie drove their big, gas guzzling 2002 car, John sat on the passenger side. He picked up the book from the floor of the car and thumbed at the pages. Pictures in the book showed objects whose far side somehow looked closer than their near side—impossible topology distortions. He shook his head and lay the book down.

John squinted through the bug flecked windshield at Parking Structure Three. The top level, the fourth deck of the structure, seemed fuzzy. He needed to visit his optometrist again. It was hell getting old. He was fifty two years old and the fringe of hair that remained on his head was graying.

John wiped the sweat off his forehead in annoyance—it was a July afternoon. Heat waves shimmered from the pavement. A dry east wind sprang up out of nowhere and blew an assortment of old newspapers and dead oak leaves, skittering and scraping over the blacktop, up against the parking structure. The wind died, releasing its burden. The debris settled in the white flowered bushes growing along the concrete walls of the structure. Millie stopped the car just short of the entrance.

John rubbed his hand over his bald head. He shuffled his feet on the car floor in agitation. He cracked his knuckles loudly and chewed at his grey mustache. "Well, what are you waiting for? Are you going to drive in there or are you going to sit here at the entrance all evening?"

His wife Millie drew in a deep breath and sighed. "Be patient. I have to wait for the entry arm to rise and let us in. You asked me to drive today. Want to take over now?"

John said, "We're gonna be late for your OCD lecture. Those useless quack psychologists."

Millie put the car in parking gear and sat there. She folded her arms, and turned to glare at John. She shook her gray haired head sadly. "Not quacks," she said. "I'm the one who has to put up with you.."

"Yes—quacks, by God," John said, "And I'm not compulsive. Things just gotta be logical and orderly around me."

John was a Harvard Law graduate. He knew better than everyone else. He distracted himself from his agitation by using one finger to trace triangular patterns on the back of his other hand, connecting the freckles and age spots with nice neat imaginary lines. All the lines formed tidy right triangles, straight out of a Euclidean Geometry text.

Millie simply sat there behind the steering wheel waiting. One at a time, several cars pulled in behind them and the drivers began to honk their horns. A real racket. John's fingers began to twitch.

He pointed at a hand-drawn sign. "Look—it says, *For Lecture, Living With OCD, push entry button and go to level five.* Did you even push the damned button yet?"

The whir of Millie's electric window was the only answer. She reached out, pushed the button and the entry arm immediately rose to admit the car. John smirked. He sat in his seat grumbling under his breath about logic, common sense, parking lot buttons, the damned people who designed them, and idiots who honked their horns.

The car jerked to a sudden stop, the tires screeched on the smooth concrete surface, and John bumped his forehead on the dash.

"Ouch. Sonofa... Why?" He rubbed his forehead and turned to Millie.

Millie pointed at another sign, *To Go Up, First Go Down Two Levels.*

John said, "Just do the deal."

"But we need to go up to level five, not down."

John scowled at her and said through clenched teeth, "Do it."

The next level down was marked *Level Minus One.* John said, "They ought to hang the engineer who designed this structure."

He looked around. All the cars on this level were old. Vintage about 1980. Kept in real good condition though. Polished, shiny, mint condition. John looked at Millie. He was being an old S.O.B. Too hard on her. She couldn't help the fact that she wasn't logical like him. She didn't even have a law degree.

He looked lovingly at her young face—the smooth face of a nice forty year old, not the wrinkled face one would expect on a fifty year old. He eyed her black hair. She must have dyed it again, and he hadn't even noticed until now.

"I love you. You look real pretty tonight," he said.

"Humph," she answered.

The next level down was marked *Minus Two.* The air smelled damp and moldy. The parking spaces were filled with classic cars, all from the era of about 1950 to 1960, and each vehicle was in amazingly good condition. There was some real money here.

"Wow," he said, "Is there a car show here today?"

John squinted out the window into the darkness. There, a dozen yards short of the next turn, his parents' old 1956 pickup was parked in one of the spaces reserved for the handicapped. He recognized the license plate. A stooped old man stood next to the car. It was John's dead father.

His father looked up, caught John's eyes, and darted into a doorway leading down a long corridor that receded into blackness.

"Wait. Stop," John said. He yelled at Millie. "That was my Dad. Dad was right there by his car."

"Oh John, John, John," said Millie. "Your father has been dead ten years. Your mother too. And their old car was hauled away to the junkyard and sold for spare parts."

"No," John insisted, "It was him. Him." John's breath came in fast shallow gasps.

Millie glowered at him. "Whatever is wrong with you tonight? Have you lost your mind?"

John looked out the car window again. He thought, *She's right. Dad is dead. Maybe I'm losing my mind, or at least my sense of logic.* Besides, they were 100 yards underground on this so called Minus Two Level. Where would a doorway and corridor lead to? He wiped his wet forehead with his shirt sleeves.

"Here we go," Millie said, "Here's the up ramp to the higher levels. Now we're getting somewhere." She sounded young, musical.

John turned to look at Millie. She looked like a teenager. Or was it just the darkness of this level?

Several bends later, they passed a sign which said, *Plus Two Level.* All the cars here were ultra-modern looking, almost weird. Stubby electric vehicles, low slung cars with ergonomic shapes. Some even had wings. There must indeed be some sort of car show going on. Maybe they could stop by the show after Millie made him sit through the lecture on obsessive-compulsive disorders. He would at least humor her.

He looked at his watch. It was already 7:15 PM. The date was wrong though. It said, Tuesday, 2011, and it was really only 2001. Stupid watch. He hated defective merchandise. Another trip to the damned shop. His fingers began to twitch again—he rubbed them along the seams of the fuzzy acrylic wool seat covers.

"We've already missed the first fifteen minutes of your useless lecture," John said.

The car slowed again, then stopped. "Something is wrong," Millie said. "We're on level four, and there are no cars here anymore. And I remember the sign said this is only a four level structure. So how can we park on level five like it says we should?" Her voice sounded tired, defeated. Millie's hair was grey again. Her face was wrinkled leather. The light in this odd parking structure must have been playing tricks on John's eyes. Then he saw the tears on her face.

A pang of guilt washed through him. "Let me take over driving so we can make some sort of progress," he said, getting out of the car and walking around to the driver side. As John settled into the driver's seat, and put his hands on the sweaty plastic of the steering wheel, Millie slid over. Her seat clacked as she ratcheted it down into a lower position. She groaned softly with arthritic pain.

He said, "I'm sorry I was so cranky. I'll try to do better." As a mumbled afterthought, he added, "I love ya."

Millie lay back on the reclined seat. "Sorry? That's what you always say. But you never change. Everything changes around you, but you're always the exact same person." She turned away to stare out the window.

John drove round a corner marked *Up*. They were still on level four, they needed level five, so going up seemed logical, in fact obvious. A few minutes later, they rounded the same corner—still on level four.

"What the Hell is wrong with this structure?" John said. He raced the engine and roared the car around several bends, deliberately going the wrong way at the next corner, no matter what the sign said. A new sign said, *Level six*. But they needed level five, not six. How had they skipped it?

He turned and shouted at Millie. "You and your stupid psychology lecture." Millie cringed against the far window. John looked out through the open sides of the structure, hoping to catch a glimpse of the University buildings to re-orient himself. What he saw caused him to slam on the brakes. He flung open the door, jumped out of the car and ran to the side of the empty sixth level. He grabbed at the cold concrete column as he leaned forward and looked out over the countryside.

It should have been almost sunset this July evening, but the sun was nowhere to be seen. It was somewhere behind a slate grey sky which hung low over the city of darkened buildings. There were no cars on the streets.

Mounds of snow lay on the concrete sides of the parking structure. An icy wind blew snowflakes onto fluffy piles and onto his shirt. This couldn't be happening anywhere in July—especially in Los Angeles. He walked back to the car and climbed in. He stared blankly at the steering wheel, breathing hard.

"What?" asked tired old Millie.

"It's snowing. In L.A. In July. And we're on a sixth level that doesn't even exist."

Millie said, "See what I mean about your trying to force your logic on everything?"

John fists tightened on the steering wheel again. "I'm gonna back the car right out of this damned place." He revved the engine and craned his neck painfully out the window, backing around a turn. He screeched the car to a stop again, just in time to avoid backing over the sharp two inch spikes lying in the car's path. *One Way, Do Not Enter, Severe Tire Damage,* the sign said. The only possible direction to go was up.

"Those god damned engineers," John said. He pounded his fists on the steering wheel and shouted. "This place was built by idiots and lunatics. Isn't there any way out?"

Millie said, "Maybe it's like the entry. Maybe we have to go up first, just to get to the turning point for the down-ramp." She looked at the snow, shivered, and turned on the car heater. She held her bony old arms around her while her teeth chattered. Some of those teeth were missing.

"Hey—that's good thinking," John said in surprise. He threw the shift lever into forward and screeched around the corners. As they passed *Level Seven,* John looked over at Millie. She lay in a huddled heap.

"Millie. Millie. Wake up—are you okay?" He slammed on the brakes and reached for her. He shook her shoulder, and she flopped over onto the floor of the car, face up. It was the white boned face of a broken-toothed skeleton which stared sightlessly at him from a tangle of grey hair.

John screamed in terror. He gunned the engine and raced, tires howling, up through the next three levels, weaving and darting at random, nearly spinning out of control on the ice slickened concrete surface of the parking structure. He had to reach the turning point and get back down to the surface, call 911, and then maybe all these crazy things would be okay, and Millie would be Millie again.

John drove on, zipping through level after level. Finally, as he passed under the sign that said, *Level 449,* he thought, *this is the second time I've passed the 449th level. God—we're thousands of feet up in the air. Four level structure indeed.* John's ears popped with the altitude. A frozen seagull lay off to the side, by the wall, with its stiff feet in the air. He remembered passing it a minute earlier.

A minute later, he passed the gull again, even though he could swear he was still driving uphill steadily. He was still on level 449. "This must be the top level at last," he said loudly, cheerfully, pretending that somehow Millie could hear him.

But top level or not, there was still no down ramp. They were still stuck, he realized, as they passed the gull a third time. He stopped the car and then backed down around the turn until he reached another set of the one way tire spikes. He stopped and set the hand brake. John climbed out of the car and ran back around one bend after another, heading downward, searching, slipping on the ice, shivering, waving his arms in the sub zero cold winds that moaned at him. If only he could find there was some way down, then he would drive the wrong way in the car—who cares about tire damage. His heart pounded in his chest.

In a minute, he again came upon the gull, this time from the other direction. For some crazy reason, when you were at 449, there was no way up and there was no way down.

John shouted into the freezing gale. "Stupid car." The words echoed once on the concrete walls, and then were lost, absorbed by the white mounds of soft snow which lay along the sides of the drive path. He kicked the frozen gull, sending it sailing through the air, where it crunched against a cement post and tumbled to the iced pavement. The gull lay still while it was covered by a settling cloud of its own feathers. John turned and walked slowly back to the car, head hung low.

When he reached the vehicle, he saw that all four tires were flat. He groaned.

"Why, why?" he said. He hadn't even driven over the one way spikes yet. He pulled at the door and it came off the rusted hinges right into his hand. The car had been brand new just a few hours ago. The door dropped clattering onto the concrete. Millie's clothes lay in tattered pieces on the floor of the car, exposing her bare white skeleton to the cold winds.

Seeing Millie like this made John quake. He gasped for breath and reached out for her with a trembling hand, just to touch something that was somehow, someway, her, and to reassure himself that he was still sane.

"I'm sorry, Millie. I'm so awful sorry." He stroked at her shoulder bones and her skeleton crumbled into chunky grey ashes right before his eyes, wafting little puffs of acrid dust into the cold air.

John screamed "Noooo." He sneezed. He ran over to the side of the structure and looked out over what had been the city. There were no buildings left. Here and there, a few steel girders and tumbled concrete blocks poked up through the otherwise uninterrupted expanse of a white ice field. A tired red sun sat low above the horizon.

John shook his head. He looked at his watch. It read 9378. He reached up and felt at his face, searching for wrinkles and missing teeth. His face felt just like it always did. He didn't even have a stubble of whiskers ... because he had shaved just this morning, or nearly 7,000 years ago, depending on how he looked at it. He thought of what Millie said, it seemed like just a few minutes ago... *everything else changes ... You never change ... reality wins every time.*

John took one final look at the forlorn sight, and then he climbed to the top of the concrete wall overlooking the frozen expanses of rusted girders. He swayed back and forth in the gusts of frigid wind. While he watched, the sun started to sink over the horizon and a misty grey began to eat away at the scene, as if the colors were somehow draining off, leaving an old black and white motion picture in place of reality. He looked at his watch. It had stopped.

John whispered, "You were right Millie, you were right all along." He closed his eyes. He relaxed and let himself fall forward into the timeless grey nothingness, letting it take him.

The Cyclotoon Perspective

Neal Skorpen

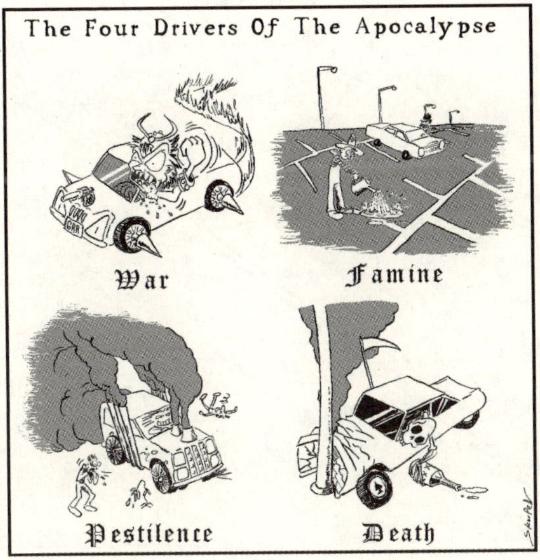

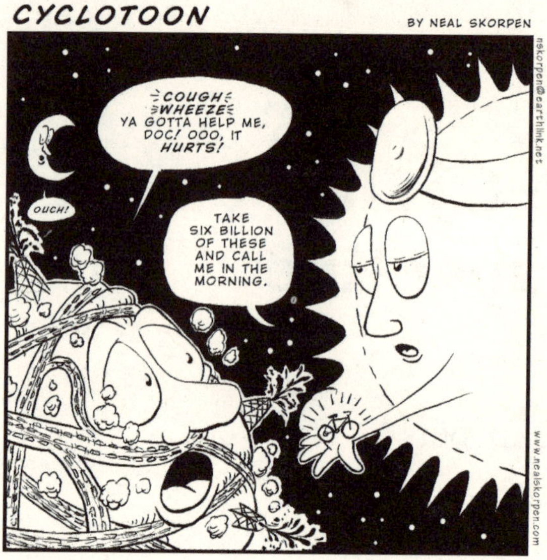
CYCLOTOON
BY NEAL SKORPEN
nskorpen@earthlink.net
COUGH WHEEZE YA GOTTA HELP ME, DOC! OOO, IT HURTS!
OUCH!
TAKE SIX BILLION OF THESE AND CALL ME IN THE MORNING.
www.nealskorpen.com

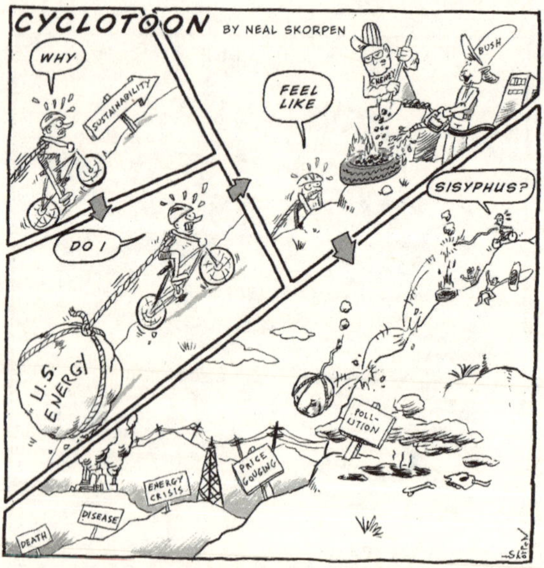
CYCLOTOON
BY NEAL SKORPEN
WHY
SUSTAINABILITY
DO I
FEEL LIKE
CHENEY
BUSH
SISYPHUS?
U.S. ENERGY
POLLUTION
PRICE GOUGING
ENERGY CRISIS
DISEASE
DEATH

CYCLOTOON BY NEAL SKORPEN

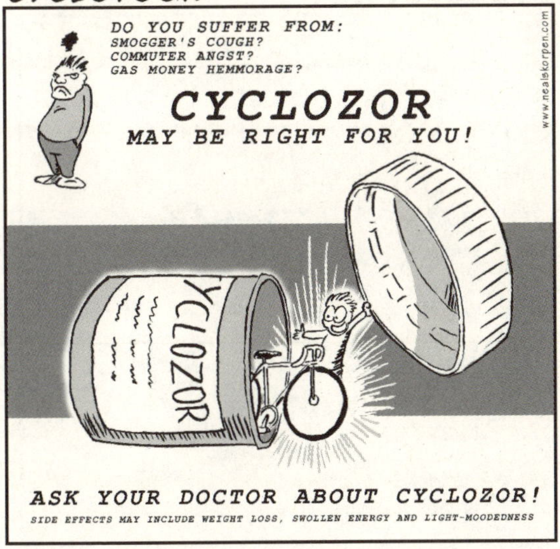

CYCLOTOON BY NEAL SKORPEN

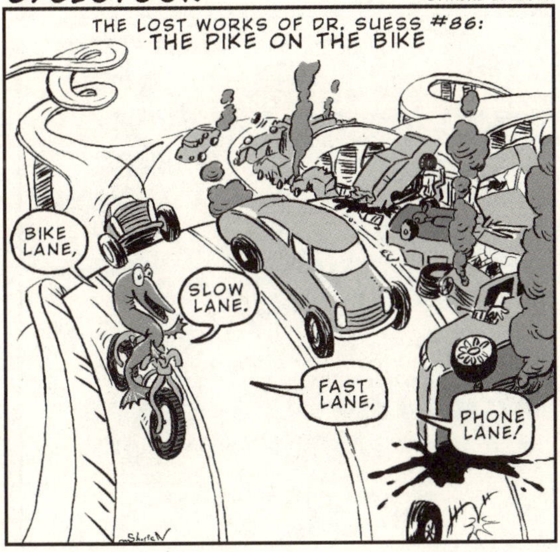

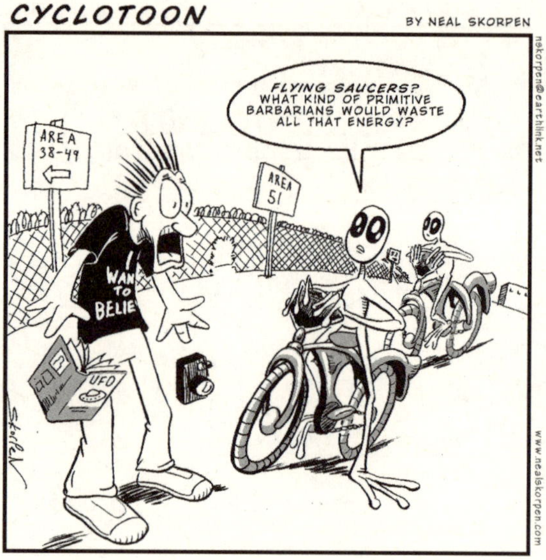
CYCLOTOON
BY NEAL SKORPEN
nskorpen@earthlink.net
FLYING SAUCERS? WHAT KIND OF PRIMITIVE BARBARIANS WOULD WASTE ALL THAT ENERGY?
AREA 38-49
AREA 51
UFO
www.nealskorpen.com

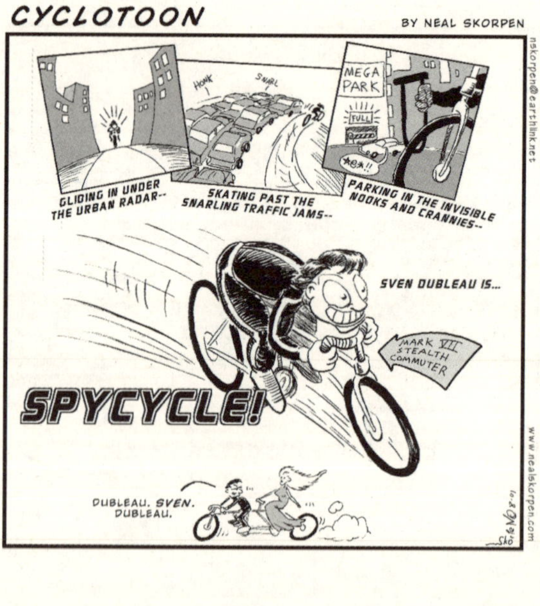
CYCLOTOON
BY NEAL SKORPEN
nskorpen@earthlink.net
HONK
SNARL
MEGA PARK
FULL
GLIDING IN UNDER THE URBAN RADAR--
SKATING PAST THE SNARLING TRAFFIC JAMS--
PARKING IN THE INVISIBLE NOOKS AND CRANNIES--
SVEN DUBLEAU IS...
MARK VII STEALTH COMMUTER
SPYCYCLE!
DUBLEAU. SVEN. DUBLEAU.
www.nealskorpen.com

Goulais River

Matt Hern

July 16th, 2001. North of Sault St. Marie, South of Wawa. Highway 1.

It's the middle of July and I'm driving through Northern Ontario. Alone. In a machine I am developing a deep antagonism to. Let's not go too deeply into the reasons why I'm here. They're foolish and relatively embarrassing. I just want to say that a series of bad decisions has led to this situation, a circumstance that I would really rather not be in. Suffice to say that I hate this goddam car, and I'm pretty sure it feels a similar loathing for me.

B.C. is a long ways away. Too far to really comprehend right now. I try to measure the distance on the map with my finger and count out days. Six? Too long. Its already been two and a half since I left Montreal. Check the distance on the scale. What? Sixty-five hours driving time? Don't think about it. Just drive. Get Zen. Be in the moment.

But again: there's that burning smell. Am I imagining it? Please let it be a flashback. Nope. It really is that burning smell again. Pull over. What the hell can it be making smoke billow from the left front tire? The whole rim has heated up alarmingly. I checked and repacked the bearings yesterday. They were fine. The shoes are in place. The pads look alright. What is going on?

This same problem stalled me yesterday, and after spending four hours in the Soo rassling with the bearings and brakes (and finding nothing amiss), I was hoping against hope that I had accidentally fixed the thing. Before that it was a fan belt in Sturgeon Falls. Before that the solenoid went, in Pembroke. The starter still is temperamental. The fuel filter needs to be cleared every once in a while. And still this burning left wheel.

I wedge the car as far onto the shoulder as possible, flag down a passing SUV, use their cell to call a tow truck and slump in the passenger seat. Through the driver's window I have a nice view of Lake Superior. Its actually kind of pretty in a Northern Ontario kind of way. The highway is quiet, only two lanes out here. I get out and walk around, pick a few blackberries, listen to the crickets humming, think about bears. My foul mood slowly lifts and I crash out in the grass on the other side of the road, facing the lake.

Chet wakes me up about an hour and a half later. Standing on the back of his flatbed tow truck, hollering for "Whoever called me to pick up this piece of shit". I swing out of my daze, stagger out of the grass and cross the road to shake his hand. I like Chet immediately. He's about 20 years old, is clutching a tall can, has no shirt on, and flicks the long bangs of his mullet out of his eyes compulsively. He's wearing tight GWG's and steel toes, and has a good glint to his eye. "B.C., eh? Got a ways to go. (Chuckles, looking at my vehicle) Good fucking luck in that thing. (Turns back me to me, grinning) Some good dope out there, huh?"

A few minutes later we're hauling ass back to Goulais River. Chet drives his rig like a madman, careening along the highway as it twists and turns the 40 k or so along the Superior shore to his shop. Cindy (his girlfriend) and I are thrown around the cab like dice in a hand while Chet pounds beers and hoots at me when I lock my door and fish for a non-existent seat belt.

Calling Goulais River a town is seriously pushing it. I can't even see the rough outlines of a village or community, but Chet swears its there. "The Goolie is just the Goolie, man. We know we're here." Arriving at his house/garage/

parts yard, he leaps out of the cab, pops another brew and gets on the cell. In under five minutes there are folks showing up from every direction. Billy comes screaming in from up the road on a dirt bike, beer in one hand, massive joint in the other. Kojak walks over from his place across the highway. Moon Dog arrives in his jacked up 4 x 4. Wes, Jeff, Candy, Amelie, Andre and Little Bill all come out of the house or shop. To a person they are sweet, welcoming and commiserative. Each shakes my hand, pats my back or ruffles my hair and offers me a beer. "Don't worry about yer car. Chet'll fix you up. Sit down, have a cold one and relax."

Soon, after a couple beers, I do in fact chill out, sit down in a lawn chair on the driveway and watch a couple of the lads tear into my vehicle. There are three huge rottweilers sitting with me, assessing the work being done. Chet doesn't even bother to take my car off the truck, but already has the tire removed and bearings in his hand. After some consultation with the assembled crew, he turns round: "Well, I don't know what the fuck the problem is. Must be some bad grease in the bearings. Let's repack em, clean the pads and the well out good and send you on your way. Alright?"

Alright. Except its getting dark, I have three thousand miles to go, and my car is still totally unreliable. It is also getting expensive. And I am not getting anywhere fast. In fact, this trip is costing me vastly more than the train or bus, and I am way behind schedule, with dubious prospects in front of me. What was I thinking?

Chet is a good guy, and doesn't charge me anything. He also invites me inside to spend the night, which quickly devolves into a long evening of drinking blackberry hooch with a bunch of off-season loggers, playing cards and shooting .22's at beer cans in the yard. By midnight I am truly gooned, and actually listening to their job offers: "C'mon, come work with us. We need another guy to drive the skidder here for a coupla months. The mill just filed for bankruptcy, so we can't sell anything, but it's good work out at the bush camp. C'mon?"

By three I pass out on the couch, jostling for space with the dogs, my head a total mess. The prospect of tomorrow: driving through Northern Ontario, almost certainly with a killer hangover and a vehicle that might well fail me at any time, haunts my dreams.

Considering everything though, I sleep more or less soundly for four hours until the house rouses. All around me are people hustling here and there, cheerful, bacon frying, dogs growling, jokes, laughing, teasing ... and I'm a wreck. Don't these people get hangovers?

Fade the rest of that day to black and return to Chet's yard. Nine hours later. Same people, same place. Same fucking problem.

In short: I got up, pulled myself together and drove off. Things went smoothly for a couple of hours until that same burning smell returned, and an almost carbon-copy of the day before ensued. I pull over. Call Chet. He and Cindy come get me. We go back to the house, drink beer and start trying to figure it out.

So here I am, sitting in the same lawn chair, drinking with the dogs again, talking to Amelie about the Leafs while a couple of lads try to figure out what my problem is. (Aside from the fact that I am an idiot for buying an ancient muscle car, trying to drive it across the country, and that I am currently missing work and pissing my family off). Kojak figures it has to be that the shoes are rubbing against the wheel well, and eventually a consensus emerges: it's the brake shoes. We can either get a pair shipped tomorrow from the Soo, or maybe someone will drive me into town to get some. One way or another, I'm not going anywhere West tomorrow, and this situation is now definitely going to cost me a coupla hundred, even though Chet says he'll only charge me for the tow again.

The evening unfolds pretty much the same as the one before. Like Jeff says: "We drink anyways, but its sure good having a visitor to party with." This time the watermelon wine comes out too, and we get more guns and Little Bill sets up the targets in the woods. Then late, we all troop across Highway 1 to Kojak's place, to shoot pool and test his 'wine' out.

The evening is a good one, but this time I don't sleep well. Not even close. I go to bed in Kojak's basement guest room, and its already getting light out. The watermelon stuff is not sitting well with me. I want to chuck, my head is ruined, I'm freezing, I have no blankets, I can't stomach the moose that Cindy fed us for dinner, I need water, there are mosquitoes everywhere. I'm a mess.

I don't actually sleep at all. I enter that passed-out too-drunk semi-sleep state for a few hours, and my thoughts turn very evil. I am full of rebuke and self-recrimination. I hate my car. I want retribution. Why did I buy it? Why not take a train? Now I am losing money and time and patience in big chunks.

I lie in misery, my car, the traffic and the highway relentless in my head. I'm hating cars, the noisy, polluting, expensive motherfuckers. The greenhouse effect. Global warming. 650 million cars world-wide emitting four billion tons of carbon dioxide per year. And the bastards are reproducing like rabbits. A new car every second. 60 cars a minute, 3,600 cars an hour, 100,000 cars a day. The number of new cars is increasing by about 38 million a year. Pollution, noise, money, urban disaster, death? the bastards are coming for us. The cars are winning the battle and we're not even fighting. I, in particular, am losing badly here.

Now I'm up. Full of righteous anger. I hate my pathetic little 'muscle' car even more, but I am propelled by righteousness, I'm on fire with justification. Someone is going to pay. Something is going to pay. For my predicament. For this whole mess. If I don't start turning the tide, who is?

I reel out of Kojak's basement, tear on some pants and my boots, can't find a shirt, no matter. The early Saturday morning sun dazzles me, but its only six-thirty and still cool in the Goolie. Staggering through the carport I reach into Kojak's truck and grab his wrap-around sunglasses.

The highway is empty. Chet's house is deadly quiet. Either everyone is asleep, or maybe out fishing already. Whatever. I'm on fire, and my prey sits on its own, within reach. My car is on the grass, about thirty yards from the house. The sight of it, still jacked up, the fuzzy fucking dice dangling from the mirror, the chrome bumpers, the racing stripes, fills me with rage: violence fills my mind.

Grimacing absurdly, full of a hazy sort of elation and adrenaline, I sprint over to a junk pile of old parts and grab a rusty tire iron, spin back, and begin attacking the car. I'm a madman: smashing it, swinging the metal hard, shuffling around the car, busting every window, driving big dents into the body, caving in the trunk. After two full rotations I tire, my effectiveness wanes, and I chuck the iron aside. I am covered in grime and sweat, I have cut my torso twice and there is a little trickle of blood on my chest.

I take a half-dozen steps back, consider (if I am capable of really considering anything) and reassess. Suddenly, the rage wells up again and I charge the car. I plow into it like a linebacker snuffing a screen. I crash into the frame, the car refuses to budge, and I recoil with glass shards riddling my shoulder, momentarily sobered up.

The sight of more blood underlines the reality that this machine has declared itself an enemy. Recovering my energy I run over to the garage and grab an armload of greasy newspapers. Tearing and crumpling them up into strips and balls, I fill the driver's side floor up and jerk a lighter out of my pocket. It takes several stabs, but eventually a dull flame catches and soon the front of the car is burning. Actually smoking. The paper is packed too tight, is too damp or something and the plastic on the seat just smolders weakly, emitting a huge volume of acrid, black smoke. I lean into the mess, trying to fan the flames, and get only mouthfuls of smoke and fumes, half asphyxiating myself, and forcing me to take a knee.

Gasping for air with that watermelon hooch just barely staying down, I retreat, daunted. How the hell do people destroy cars without crashing them? All of a sudden a vision of dust flashes through me, and overtaken by another rush of pleasure and adrenaline I sprint to the house and grab a shotgun sitting behind a couch on the back porch. Staggering under the weight of anticipation, I reel out, loading a shell with three more in hand.

I stop twenty yards away and glare down the barrel. Now the car looks pathetic and small, like a baby bunny. Cowering.

I have fired guns before. And enjoyed it every time. But this is the first time I have used a shotgun, and I don't prepare for the kick. It really does feel like getting kicked, by a mule maybe. With my shoulder and neck throbbing, focusing my attention acutely, I contemplate the trajectory of my shot. It rips through the broken shards of the back window and imbeds somewhere in the dash with a sullen crash.

The quiet after a gunshot is a beautiful thing. The boom echoes all through the trees and across the highway, and in its absence is a certain kind of real clarity. The car does nothing though. It absorbs the shell and stares back, daring me to fire again.

Why has no one come out of the house? They can't still be sleeping. I just attacked a car with a tire iron, tried to burn it, and then fired a shotgun, all in their front yard. And no one has come out to ask me what the hell I am doing?

I am too excited. I will myself to cool down and think. This works on screen. This has to work. Maybe I need a grenade or an anti-aircraft thing or something. They probably have an elephant gun in the house. But wait. The gas tank. Oh yeah. Yeah. That's it. It hits me like the slow dawning of an ecstasy rush. Thrilled, I take a few more steps back and move around to the side and find the place where you put gas in. Assuming the actual gas tank must be around there, maybe a little lower, I cock the gun again, and fire, this time readying for the kick.

It's perfect. Just like a cheesy action movie. The slug disappears, almost in slow motion, and then in a welling ball there is a series of small, then large, then awesome explosions, and within forty seconds the car is totally engulfed in flame. There are deep yellow and orange flames, licking ten feet up, and a mountain of black smoke pouring upwards. I am stunned with my own genius and the beauty of firearms.

And still, no one is coming out of the house. Maybe explosions in their front yard aren't enough to disturb a weekend of sleeping in. Why is no one coming out? Not even the dogs. Ah well, their loss. Why are there no cops coming, no sirens, no outraged neighbours? Maybe some psycho blowing up his car on a Saturday morning is standard Goolie River fare.

I step back, covered in sweat and grime. Nausea floods over me. I lean over and retch long and hard, but nothing comes, aside from a long strand of yellow mucous. I stand up cautiously, fish in my pocket for the three ibuprofens that I had stored there in anticipation, and chew them carefully. The flames are hot, the smoke black, the sound snap-crackle-popping. The whole thing is almost too perfect. And still no one comes. Where the hell is everybody? Am I the only witness to my own little act of god?

I sit quietly for thirty more minutes, eating an apple picked from a tree in the yard, feeling the satisfaction of the drugs take hold and my nausea recede enough to keep the fruit down. Like my hangover, the flames are dying down and I can see parts of the car clearly; the seats, the dash, the roof, all of it blackened and maimed, still burning hot, the paint and plastic all gone, just the frame sizzling. My satisfaction is huge, and is tinted with the shiny sheen of revenge.

I check one last time that the flames aren't spreading onto the grass, carefully put the gun back in the hallway, shoulder my bag and walk out onto Highway 1, heading east, mentally assessing how long it will take me to hitchhike to the bus depot in Sault St. Marie.

Three Poems from *Asphalt*

Ray McManus

Gridlock

It's hot and you close your eyes
You close your eyes and sink
going nowhere, not wanting to go
anywhere, wanting to go somewhere,
not thinking about going, not thinking.
You pull down the visor. You bite
the inside of your cheek, take your tongue,
and hold it there. You see faces, old faces,
faces like yours, ugly and pale, faces
that look hungry, look angry, look tired,
tired of moving, tired of not moving,
until the stillness breaks and you
inch your way closer to corners, to stop
lights, to store fronts, and windows, skin
to vinyl, piston to engine, transmission to axle.

And it's hot man, my God, it's hot, fumes
rise from the blacktop, men on the corner
sweat, little girls on the corner sweat

harder, louder, jump on one foot and scream.
You mustn't stare, try not to stare, can't stare,
stare anyway, at the little one not jumping,
the little one with finger in mouth making sounds
because she has no one to talk to. The little one
reminds you of your sister, your little sister,
your fat sister, your pain-in-the-ass sister,
the sister that calls too late, the sister that grew
too fast, the sister that you don't talk to anymore.

Pavement

By now you should be thinking
about what this really means

You should be thinking
about where this could be

going, about where you could
be going. Without headlights,

lap belt, and wheel you will die,
shot out of the window, thrown

through the windshield,
into darkness, onto pavement,

gravel, dirt. That's it. Instead
you look down, shift your foot

against the curb, push off, open
your eyes. The force is not enough

to hold you there. You are centered
and rooted, but your head is light,

all you see is a bumper sticker
that reads *Whatever*, asphalt reflected

in chrome, a plastic dog on the back-
dash of the car that just left you.

Main Street at Eighty

is one big blur, a sunny eastside,
before the last street, curb, post, and walk,
the last stroller, squeak, and scream,
before you forget about the force around
your waist and neck, the planting of foot
and seat, the shift itself, everything moving
too fast to pin down, and you have
only scuffed the surface: hotdog vender,
elm tree, street sweeper, speed bump,
then car wreck, hemorrhage—what's in, out,
gone. And it should mean more than crack, block,
or cavity, more than old bolts and sockets,
lungs and rearview mirrors, what comes
out of them, what goes in—asphalt,
gas, and stain, something slick, soft—
a small blue sandal in the middle of the intersection.

Documentation: The Right Of Way Manifesto

Right Of Way

Poe's story "The Purloined Letter" turns on the idea that sometimes the best way to hide something is to leave it in plain sight. The idea is amply illustrated for New Yorkers by two glaring scandals that can be seen in action, any day, on any streetcorner, by anybody who ventures outside his house:

- There's an unacknowledged crime wave of lawless behavior by drivers—a crime wave that takes the lives of 250 or so New Yorkers every year.

- In a gigantic theft of public property, the car-owning minority of New Yorkers has stolen the streets from the non-motorized rest of us.

Thugarchy, the Rule of the Road

Stand at a busy streetcorner in midtown any weekday morning and just watch what happens. You'll see a dozen dangerous crimes a minute: drivers "squeezing" the light or just plain ignoring it, drivers bullying pedestrians out of the crosswalk in a very lopsided game of "chicken," drivers stomping on their accelerators and peeling out of a stalled lane into another that offers an irresistible ten feet of Open Road, only to shudder to a squealing stop half a second later.

Enforcement is essentially nonexistent: when was the last time you saw, or heard of, a driver being ticketed for not yielding the right-of-way to a pedestrian? Or for reckless driving after forcing a cyclist off the road? It just doesn't happen.

Speed limits, of course, are a standing joke; most drivers in New York couldn't even tell you what the speed limit is. Whenever there's an opportunity for drivers to open it up and make time, God help the carless. And quite apart from the nominal limit, the concept of "reckless driving" simply has no meaning at all to drivers and police alike: cars roar down narrow side streets at thirty miles an hour, desperate to make a light; they may, by chance, be under the limit, but they're way above what is safe and prudent.

Obviously, nobody cares: not the Mayor, not the cops, not the District Attorneys. The police have an informal, though freely acknowledged, rule that a driver who kills a pedestrian has to be violating at least two laws before he'll be charged with anything. So the vast majority of killer drivers roll away from the scene of the crime without even any points on their license.

Not only do the authorities not care, but most New Yorkers simply accept these conditions, without indignation, as an inescapable fact of life; the Purloined Letter principle at work.

But this tyranny of motorized criminals is not a law of nature. It could be otherwise. It will be otherwise when people wake up to the reality of what is being done to them.

So the first of our goals is this: to make the invisible visible.

Who Owns the Streets?

Everybody knows the theoretical answer: the streets belong to all New Yorkers. But reality is quite a different matter.

The reality is that the Department of Transportation has one priority: to move as many cars as possible, as fast as possible. All other street users—pedestrians, cyclists, skaters, people in wheelchairs or children in strollers—are an afterthought at best.

Again: try the experiment for yourself. Go to a big, busy intersection and time how long the "walk" lights last. Go to 95th and Riverside, where people crossing from one section of Riverside Park to the other are expected to walk around three sides of a square, crossing four streets, in order to facilitate the lordly progress of cars to and from the West Side Highway. Or look around in your own neighborhood: you're sure to find a similar example of through-the-windshield engineering.

Consider former mayor Giuliani's pedestrian barricades in midtown Manhattan, erected to prevent pedestrians from crossing the one-way avenues on the "downstream" side of intersections. Instead, we are expected to cross three streets instead of one, and wait for three lights, in order to avoid delaying drivers impatient to turn from cross streets onto the avenues. These cage-like structures, with their architectural rhetoric of incarceration, are the sort of thing we came to expect from Giuliani; but his successor, the supposedly enlightened technocrat Mike Bloomberg, has also refused to remove them.

Think how much public space is devoted to free parking. Everybody knows how expensive New York real estate is, yet we reserve a huge amount of it for drivers to leave their cars in, and we don't charge them a penny. How much would you have to pay for another hundred square feet of space in your apartment? That's what a car takes up at curbside.

All this in spite of the fact that car owners are, by a substantial margin, a minority of New Yorkers: around 40% of households (not individuals!) city-wide, considerably less in Brooklyn and Manhattan.

But car owners, like gun owners, are regarded by politicians as a constituency. The carless are not so regarded, because they have—so far!—no consciousness of their status as a disenfranchised majority.

This, then, is the second of our goals: to awaken the sleeping giant—the majority of New Yorkers whose civic patrimony has been stolen from them for the benefit of a greedy few.

What do we want?

Bringing cars under control is not an insoluble problem. Cities elsewhere have done it. There's a well-understood set of techniques for the purpose.

- Enforcement. Easy, and obvious, and it pays for itself in fines. Are you in the intersection after the light changes? You get a ticket. Brush a pedestrian out of the crosswalk? Get a ticket. Driving faster than is safe on a narrow side street? Get a ticket. A few hundred dollars per driver later, our drivers will be the envy of the world for their tameness and civility.
- Reduce the flow. One reason drivers are so hysterical in New York is that there are just too many of them trying to fit into too little space. Sure, there are people who really have a legitimate reason to be driving. But there are plenty who don't. There are lots of ways to sort out the two groups. One technique is simply to charge drivers what they really cost: including their costs in land use for roads and for parking, the costs of road and bridge maintenance, and the indirect costs of noise, pollution, policing. The list goes on and on.
- Traffic engineering for people, not for machines. It's time the Department of Transportation woke up to the

fact that not everybody gets around in a car. Much can be done just by changing a few settings on the traffic light controls. For example: four-way reds, so pedestrians won't be chased out of the crosswalk by turning cars. More pedestrian signal buttons, and a quicker action for them. "Walk" signals that last long enough so you don't have to be an Olympic sprinter to make it across the street.

- Education. Drivers believe they own the road—and who can blame them, since nobody has told them otherwise? Time to start telling them. Paint "YIELD TO PEDESTRIANS" in every crosswalk. Put public-service spots on TV, and make it clear in no uncertain terms that pedestrians, cyclists, and skaters are also on the street by equal right under the law, and morally by an even greater right, since their road use is less costly in every way.

The Bigger Picture

American car-madness has taken its toll throughout the country, not just in New York; indeed, in some ways New York is relatively fortunate (we still have a functioning public transportation system, for example, at least for the time being). These larger social costs include the waste of open space, the decline of community, the pollution, and of course the Vietnam's worth of deaths every few months. Who hasn't lost a friend, or more than one, in the killing grounds of the American highway system?

We can't change all this overnight. But we can make a start. And there's no better place to start than New York City, where alternatives to the car are already used, every day, by the majority of our citizens.

It can be done. And, as so often in the past, New York can show the way.

Boneshaker

Scott Munn

I ran into the Ghost of Christmas Past on the Connex Southeast into London and we got to chatting. "I miss the old trains," he sighed, an ectoplasmic Disgusted of Tunbridge Wells.

"What, steam?" I inquired politely.

"No. Slam-door. They had proper storage." He waxed nostalgic. "You could sit in the cage and keep your velocipede company if you liked. Get away from some of the mobile phones. If I have to listen to the opening bars of 'Mission Impossible' one more time I'm really going to scream."

"Don't hear that one quite so often anymore," I remarked, then considered who I was talking to. We were shoehorned into the cramped new seats ("as if people were getting SMALLER these days"), keeping an idle eye on our bikes. His was indeed of ancient manufacture, with a gaslight up front, wooden saddle ("honest English Oak, not one of your sissy softwoods"), and a braking system which probably didn't even work in theory. The only nod to modernity was a rear LED: "It's just sensible."

My ride was a sexy little number with toeclips. "And they say I live in the past," he commented.

I ignored this. "You must be in great demand. The olde days are big business. Especially in this country."

"I do all right," he admitted. "Our meeting like this wasn't an accident. I have something to show you."

"But I only bought a cheap day return." I gazed out the window. Everything was...different somehow. "I don't believe it," I gasped. "You've transported me back in time."

"No, that's just Catford," he said. "We're not there yet."

Eventually we rolled into central London and disembarked. "What year is it?!" I shouted to be heard above the clopping hooves of what sounded like a million horses.

"Let's just say black is the new black. And I'm not talking about Lycra." He glanced at my bike. "Better lose that. The Victorians aren't big on gripshifters."

We stopped by a bike shop and I traded down (up, in the shopkeeper's opinion: "What do you have against STI?") for a Boneshaker. Then we headed through town on those damn cobblestones.

"Where we going, CP?" I asked, shaken but not yet stirred.

"Nowhere in particular," he breezily admitted. "I just like riding around." We stopped near the Houses of Parliament.

"Why is it called 'Big Ben'?" I wondered aloud, just making conversation as we wolfed down some eel & chips sold by a vendor who was favouring me with a very suspicious look.

"I wouldn't worry if it was organic or not," my ethereal companion chided me. "Big Ben? That was the Queen's pet name for Albert's—" BONG! BONG! BONG!

He glanced at his pocket-watch. "Really, I do dither. What do you see?"

Not much. The smog was incredible. "A helluva lot of fertilizer," was my first thought. We'd spent much of the afternoon dodging it and the factories which produced it. "Unpredictable horses with very strong legs—and they aren't fond of sharing the road. Traffic jams and frayed tempers. An almost eerie absence of SUVs. A few brave souls wobbling around on two wheels; I guess it hasn't really caught on yet. Hats." I was mildly aggrieved. "What is it I'm supposed to be seeing? Why did you drag me back here into this godforsaken century?"

He shook his head sadly. "Young people today [presumably referring to anyone less than half an eon old]. Always bitching about something. I just wanted to show you that urban cyclists have always had a tough time. Thought maybe it would make you feel better. And I get frequent time-travelling miles."

Well, fine. "But what does all this have to do with your job description?"

He shrugged his shoulders, less than interested in debating the point. "It's Christmas. I'm a ghost. It's a gig."

I contemplated the filthy Thames. Turned around and he was gone. So were almost all the horses. Suddenly it was present day. My Boneshaker was leaning drunkenly against a wall. The eel vendor had transmogrified into a cop, who was eyeing it with official mistrust and communicating his disquiet into his radio. I rushed to retrieve it, assuring him it was not a cleverly-disguised bomb and silently cursing the ghost for not being very detail-oriented; I wanted my regular bike back.

Presumably a jobbing Ghost of Christmas Present would be along shortly. I contemplated the filthy river of cars. A courier zipped past, sucked through the metal corridor, on high alert for tourists gawking at Big Ben (if only they knew). I took my bike for a short walk along the embankment, gripping its saddle, minutely adjusting its course as if I were riding no-handed. The Ferris wheel across the Thames turned its lazy arc, spokes glinting, a monster bicycle wheel scooping up one load of riders after another for a taste of the sky. Almost like going spinning, I thought. Perhaps one day it would shake itself loose from its moorings and take them for a tour of the city, scattering taxis and delivery vans as it freewheeled from one photo-op to another.

I guided the Boneshaker into a stately U-turn around a performance artist making decent money imitating a statue and hopped on. As I glided over Westminster Bridge the Ghost of Christmas Future cruised alongside me on a tandem. The Ghost of Christmas Present was stoking feverishly and mumbling to himself.

"Ignore him," said Future smoothly. "Everyone else does." We reached the other side and all the motorised vehicles abruptly vanished. The road was awash with bicycles. Hundreds and hundreds of 'em: "The mayor is thinking of a congestion charge."

I was almost but not quite speechless. "This is impossible. How...?"

The Future smiled enigmatically, like the Mona Lisa. "I can't tell you that. You'll have to figure it out for yourselves." He increased his pace and the two of them slipped away, parting a sea of cheery bells.

Traffic Zoology

Matthew Frederick Davis Hemming

Abstract. There is a secret zoo that runs encaged along the roads.

They are liquid, semi-visible goliaths that rage through the streams and chunks of ordinary traffic, with the effervescent tendrils of mile-long tails whipping behind them like Chinese dragons. Though composed of hundreds of pounds of steel, glass and plastic, they are able to pass through solid objects. They are bound by the laws of the highway, but not by any conventional notion of time or space.

They are *Aggregate Traffic Animals*: a menagerie of emergent beasts drawn from the interacting behaviours of many individual human beings driving many individual cars with many individual goals, their collective activity giving rise to something with greater presence, power and purpose than the sum of its constituents. They take on a host of different forms, each to serve a different end.

They are real, and they drive among us.

An earlier version of this article appeared on Rusty Foster's collaborative media website at http://www.kuro5hin.org under the pen-name CheeseburgerBrown.

Preamble. In his introduction to *The Extended Phenotype* (Dawkins, 1982) enthusiastic evolutionary biology cheerleader and Commodore-hacking pop-science guru Richard Dawkins invites us to consider the Necker Cube Illusion (Necker, 1832): a two-dimensional image of interlocking lines representing a three-dimensional block in which the foreground and background can seem to flip back and forth as the brain fruitlessly seeks the "true" interpretation of the ambiguously depicted space. This is Dawkins' starting point for a thought experiment in which he blurs the lines between species, their genes and the environment, calling into question the traditional boundaries drawn through biological systems to identify the relevant level of study. To wit, to wank:

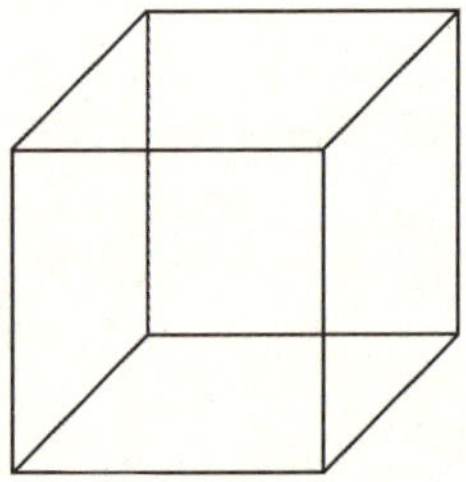

Figure 1: The Necker Cube

> We look at life and begin by seeing a collection of interacting individual organisms. We know that they contain smaller units, and we know that they are, in turn, parts of larger composite units, but we fix our gaze on the whole organisms. Then suddenly the image flips. The individual bodies are still there; they have not moved, but they seem to have gone transparent [...]

In other words, if you are able to de-emphasise the organism itself you are free to appreciate the idea of beaver ponds as artificial lakes generated by beaver genes, or to see a spider's web as an arrangement of silk drawn by DNA. By extending the lines with which we bound the traditional phenotype, we define new organisms, merging technology and individuals into communities the same way that ancient micro-organisms interacting inside bilipid membranes fell into symbiotic lockstep dances to found the first stable cells (Tien, 2000).

Organelles, cells, bodies, herds: at which level we discern the animal is purely a matter of focus.

This idea of the emergent animal or "super-organism" is hardly particular to Dawkins: William Morton Wheeler remarked on the idea in his breakthrough paper "The Ant Colony as an Organism" (Wheeler, 1911) with a treatment that is every bit as cogent but with considerably less *otaku chic* than Kevin Kelly's printed-soundbyte manifesto on hive complexity, *Out of Control* (Kelly, 1994). In the words of Kelly:

> There is nothing to be found in a beehive that is not submerged in a bee. And yet you can search a bee forever with cyclotron and fluoroscope, and you will never find a hive.

So too can you examine a driver in a car and know nothing about the greater animal in which they both participate when the circumstances are right. Some of the applicable forces can be seen most clearly in the rarified environment of the professional race course, as explored by David Ronfeldt, a senior social scientist at RAND, in his paper "Social Science at 190 MPH on NASCAR's Biggest Superspeedways" (Ronfeldt, 2002), where fleeting moments of cooperation between rivals are necessary in order to win. Ronfeldt focuses in particular on the phenomenon of draft line formation, which is similar to the way flocking birds can share aerodynamic advantage. Like iron filings in a magnetic field, the large-scale distribution of opportunistically partnering cars are drawn into predictable macro-scale patterns across the speedway:

> Once the racers sort themselves out—after ten to twenty laps—it is common to see a single draft line of four to seven cars running in front, pursued a hundred or so yards back by a second line of cars, all another hundred or so yards ahead of a large pack of cars that may still be running

> in parallel lines but are doing more dicing than drafting [...] Cars that run alone, often stuck dangerously between two draft lines, will appear to drift irrevocably backward.

Freed of the bonds of racing's formalism, the *Aggregate Traffic Animals* (ATAs) are born, rooted in transient symbioses between individual patches of drivers that will crystallise into the organs of the beast. But the circumstances have to be just right for one to emerge. The unholy Hieronymus Bosch-style concert of homicidal applied-*schadenfreude* that may characterise your urban, intra-urban or sub-urban driving experience is not ripe ground for ATA growth: too frothy.

The sociological and scatological dances of the megalopolis rushhour, too, are beyond the scope of this article, and are at any rate most likely best explored with deep computer simulations using high-tech software automata tools with average driver profiles linked to real world statistics of roadway usage coupled with a virtual army of ten thousand angry axe-wielding orcs battling an equal number of obedient clonetroopers.

Rather, this field is perfect fodder for the *amateur ethologist*, observing phenomena with a keen eye, an open mind and a sharp pencil. And, while much has been written about the wave dynamics of traffic jams and phase-transitions in traffic density, very little time has been devoted to the observation and cataloguing of persistent multi-car zoomorphia.

Early Observations. The author first became aware of the existence of ATAs while making his way through the hinterland of Canada on a long, mid-winter solo drive in a decrepit muscle-car from the seventies with no functioning radio. Due to his dangerous penchant for immersive daydreaming in the absence of external stimuli, he began to parasite his driving decisions by locking in behind another car with comparable speed ambitions. By reserving a sliver of awareness for tracking the red brake lights of the "lead" car for changes in speed or direction, the author was able to comfortably enjoy his trance while a hefty burden of road awareness was outsourced to the other driver, causing the front car to function as a sort of early warning mechanism for changing conditions (including the Mounties' speed-traps).

The notion resurfaced while the author was wrestling a rented, dented German-designed locally-built economy vehicle down a twisting, pot-holed two-lane jungle highway through the Mexican state of Quintana Roo. As the journey began he found himself hedged inside a short parade of other tourists, all driving their rental cars out of the airport at around the same time at a hesitant pace, braking frequently to process the unfamiliar leafy darkness ahead. Fearing injury, the author laterally-leapfrogged the indecisive parade and drove on into the murk alone. Remembering his success in the far north, he latched on to the back of a local vehicle (a home-modded "convertible" carrying ten people, standing room only), using its varying speed as an indicator of road conditions. Unexpectedly, this move was noticed by several of the other tourists, who began to fight to separate themselves from the melee and join the newer, more surefooted pack that was rapidly pulling ahead...

By the time the author had reached his exit the impromptu fleet of vehicles had become a persistent, homeostatic phenomenon. The fleet had learned to manipulate the spacing between its components in order to remain permeable to faster moving local traffic while defending its integrity against more disruptive external vehicles. Pulses of communication signifying when the passing lane was clear rippled down the chain through a conscientious leaning into the gravel shoulder, assisting in the process of expectorating invaders. Several of the original tourist vehicles ended up being swapped out for other vehicles without rocking the boat. Later on, even the leader was swapped out for another experienced local car.

It was a fetching game, contributing to the welfare of all of its players in an interesting way, but it was *not* a true ATA. It was too conscious a contrivance to be anything more than a delightful spontaneous social event.

You see, a distributed animal with human components can be very sensitive to perturbations from within. It is only when the conflicting threads of goals, reasoning and competition between individual human minds are quietened into the background noise that the soil can truly be ripe to raise a complex beast. When drivers can fall into a semi-hypnotic state and their herd instincts take over, the seeds are laid for something greater.

Habitat. While there are thousands of traffic animal breeding grounds along the paved networks of the world, only one driving region has been extensively explored at this time, due largely to budgetary considerations.

The TransCanada Highway is a nearly ideal environment for the production of large-scale ATA phenomena, due in great part to the simplicity of its shape: all cars are moving either westbound or eastbound, streamlining the goals of the drivers in much the same way as the shape of the Daytona superspeedway encourages drafting partnerships. Also, because there are long stretches through lonely wilderness and semi-tundra, nascent traffic animals have a long period in which to mature before coming against obstacles like influxes of new cars or navigating around towns; and because the highway wends its way directly through most of Canada's major cities, it provides a handy litmus test for the homeostatic integrity of a given specimen simply by observing whether or not it makes it through to the other side of the urban area intact.

While daytime ATA formation is not rare, it is under the cover of darkness that development can proceed in a comparatively unfettered fashion. This is due in large part to the more abstract, disconnected experience of interacting with other vehicles merely as points of coloured light. Familiar prejudices and stereotypes—potential sources of destructive competition—are smoothed out by the shadows. At least on the basis of visual impressions, a station-wagon and a sportscar can enter a system as peers.

Diminished visibility caused by mild to moderate weather conditions can have a similarly equalising effect, but when conditions become too severe, drivers tend to clump into packs for safety, leading to pseudo-ATA fleets that are all too conscious social events (as in the Quintana Roo experience).

Show me an autumn stretch of prairie transcontinental highway at twilight, and I will show you the secret zoo of the road.

Typical Morphologies. The most basic form of multi-car life is the *Asipetal Caterpillar*, also known as a *worm.* Worms begin when a stable solo vehicle spawns a linear, single-lane chain of vehicles composed of loose monomers joining at the rear (a closely related, but dysfunctional, construct known as an *Acropetal Caterpillar* grows by adding vehicles to the front of the chain, generally leading to destructive diffusion or autolysis). Short, lithe worms are the fundamental building blocks of healthy ATA tissue. Perverse, long-form worms are the seeds of congestion and death.

The second atomic element of ATA tissue stands in stark contrast to the worm, for it is a fleeting thing, and when it takes concrete form at all it is often manifested as a single car. The *Apparent Coxswain* is a vehicle that appears, to the conscious or semi-conscious mind of one or more drivers, to be a leader of the worm. When the Apparent Coxswain changes lanes, there is a higher probability that a majority of the worm will follow suit than if the change were initiated by a less trusted vehicle. In many cases each car in a worm perceives the car immediately ahead of it to be the Apparent Coxswain, leading to domino-effect lane-transitions; such formations have high homeostatic integrity because of the worm's ability to "find a new head" should one Apparent Coxswain be lost to the currents. (Please note: the *Apparent Coxswain* should not be confused with the *Virtual Coxswain* or the *Napoleonic Coxswain*, discussed below.)

Formations that achieve such integration become *Cholingers*: Asipetal Caterpillars with tightly-integrated internal feedback systems of Apparent Coxswains, capable of transmitting information from tip to tail with high fidelity. Cholingers can slither to avoid torn tyres on the road, twitch around slow-moving vehicles, and even slip through packs of alien worms, wild *axenes* and other traffic froth to arrive on the other side intact.

Of course, not all Cholingers slip through the strangers: sometimes they interact.

Every Cholinger is either *benthic* or *pelagic*. *Benthic Cholingers* travel at a similar rate to the currents of the road, while *Pelagic Cholingers* travel at a dissimilar rate when compared to other traffic (typically a faster rate). It is possible, however, for a benthic line to be *picked up* and carried along by a pelagic cousin, leading to a coupled form. This is the first real Aggregate Traffic Animal we will meet tonight: a bilaterally asymmetrical diageotrope known as the *Epiphysian Cyclosalp*.

Within the body of the Cyclosalp the individual Cholingers are transmuted into a pair of *Librigenates*—stretchy, free-flowing tissue that is bounded in space by the relationship with its partner, the accelerating pelagic lobe sliding forward and the steady benthic lobe catching up in a slow-motion slingshot, compressing and expanding between the loose, senseless clumps of other cars. This accordion-like effect might initially seem to be a force tearing the animal apart, rending pelagic from benthic—and this is indeed what might happen in too rarified an atmosphere—but when presented with obstacles of any kind, the Librigenates that comprise the Cyclosalp fall back on their Cholinger heritage of local integrity, crystallising *en masse* to navigate the hazard.

Unfettered, the Epiphysian Cyclosalp is like half a butterfly, its riparian body gilded by a slowly flapping wing of accelerating, gliding Librigenates ebbing and flowing in a stately round. Its insides whorl as partners switch places, benthic turning briefly pelagic, pacer cars joining a rippling pulse of local inertia forward, headlights cross-sweeping.

It is beyond the scope of this article to detail the myriad circumstances that provide seed for the profitable entanglement of multiple Cyclosalpic streams. So diverse are the possibilities that we could fill a Biblical tome without scratching the surface, without revealing the common thread of simplicity upon which the complexity hinges. Suffice it to say the larger clade includes such varied forms as the whiplashing *Epinastic Tricyclosalp*, the many-fingered *Dicyclosalp Fimbriatum*, and the diaphanous, fleeting wonder of the mile-long *Merosporangic Super-Cyclosalp*...

Of course, not all Asipetal Caterpillars grow up to become stately Cholingers; instead, they lock into rigid sticks of uniform properties called *Pycnoblastoids*. While short-lived *Apiculate Pycnoblastoids* (in which the Apparent Coxswain is always the most forward car) are more common, it is the more flexible *Laxiflorous Pycnoblastoid* (in which the Apparent Coxswain is any car *except* that most forward) that lives a more fruitful life.

For instance, consider the case of a typical composite entity like a *Tripycnoblastic Oomycotum*, in which independent pycnoblasts jockey for position internally directly or by proxy through one or more *Napoleonic Coxswains* (that is, drivers who suffer from the delusion that they are single-handedly responsible for steering/leading their local substructure). The domino-line behaviour of an Apiculate Pycnoblastoid makes it too brittle to survive the stresses of being permeated by a competing pycnoblast, whereas the comparatively elastic structure of the Laxifloroid—imparted due to the inherent time delay involved in co-ordinating with a mid-fleet Apparent Coxswain—retains a perfect balance of rigidity and looseness, riding a line between orchestration and dissolution that makes composite forms like Oomycota possible.

Pycnoblastic tissue is unusual in that it makes use of some level of awareness on the part of the driver that they are participating in a formation (though drivers are only likely to be aware of the local level of structure). When this awareness reaches a certain level the composite entity is usually destroyed by internal stresses, but occasionally a dissolving multi-pycnoblast will emit a stream of highly energised vehicles: the *Apheresoid Lirellate*, a concentrated apiculatoid pycnoblast flung free from the miasma of death to rocket away, using for a coxswain the abandoned carcass itself.

...These are but the fringes of the zoo, the tip of the iceberg.

We have not even touched on the sensitive antennae of the *Stipitate Phototaxites* fringed with *Virtual Coxswains*, pseudo-lead cars ready to be sacrificed to trip any trap; the chaotic wrath of the *Biflagellate Ableptic Figmo* and the fate of the cystidial flotsam locked within them; the weird rhythms of the *Cacospysic Super-Barbicanoids* and their elaborate dance of shifting coxswains; the majesty of the motorcycle-based *Raging Fallaxoid*; the menagerie of endless cancers that can grow from unexpectorated *papillic granulomae*, from cataracts of *geriatric nektons*, or from service-stations with badly planned driveways.

Further Study. The study of a new order of life is not without its risks, both professional (in terms of reputation) and practical (in terms of being maimed by mis-navigated vehicles). The amateur automotive ethologist must not only have keen skills of observation, but also the fortitude to persevere despite the slings and arrows of dubious dissenters. Like Leeuwenhoek's controversial *animalcules* and Pasteur's superstition-defying microbes, there will always exist a certain testudinal resistance to new ideas among older quarters. There will be those who doubt the very existence of aggregate vehicular life, or who insist that the zoo of the road dwells in metaphor alone.

The opinions of such sceptics could be changed by a single night spent on a grassy hill overlooking a well-travelled country highway, watching the streams of red and silver lights merge and split, compress and attenuate, roil and interact, fatten and reproduce...

Watch the roads, and see the zoo for yourself. There is no denying its patterns of insectile purpose, its myriad variations in anatomy and configuration, or the orchestrated madness of the low-cost petroleum feeding frenzy. Your own mind, honed by thousands of generations of natural selection to recognise life from non-life, will tell you it is true; the disciplines of careful observation and meticulous classification will tell you how, and why.

Open your eyes, and witness an untapped world.

Bibliography

Dawkins, Richard. *The Extended Phenotype.* Oxford: W.H. Freeman and Company Limited & Oxford University Press, 1982.

Hölldobler, Bert and Edward O. Wilson. *Journey to the Ants: A Story of Scientific Exploration.* Cambridge: Belknap Press of Harvard University Press, 1994.

Kauffman, Stuart. *At Home in the Universe: The Search for Laws of Self-Organization and Complexity.* Oxford: Oxford University Press, 1995.

Kelly, Kevin. *Out of Control: The New Biology of Machines, Social Systems and the Economic World.* Boulder: Perseus Books, 1994.

Ronfeldt, David. *Social Science at 190 MPH on NASCAR's Biggest Superspeedways.* First Monday, Vol. 5, No. 2, 2002. http://www.firstmonday.dk

Tien, H.T. and Angelica Ottova-Leitmannova. *Membrane Biophysics: As Viewed From Experimental Bilayer Lipid Membranes (Planar Lipid Bilayers and Spherical Liposomes).* Amsterdam: Elsevier, 2000.

William Morton Wheeler. *The Ant Colony as an Organism,* Journal of Morphology, Vol. 22, 1911, p. 307-325.

Necker, Louis Albert. *Observations on some remarkable phenomena seen in Switzerland; and an optical phenomenon which occurs on viewing of a crystal or geometrical solid,* The London and Edinburgh Philosophical Magazine and Journal of Science, Vol. 1, 1832, p. 329-337.

Freeway Legs

Alex Lang

In LA

S BIG S
WE R
WHEN
WE R
R CARS
TOUGH
LK WSTRN
COWBOYS

BANG!

This Spider's Work

Sitting on this stoney perch,
I watch the sun retreat
between the sea and sky

It is Friday evening
some time past five
and the sun's tail divides

the water from the sky
with tapering pink band
stretching horizontally
acting as marker
to keep wayward sea-farers
from sailing one to the other.

The soft splashing of stardust and mechanical debris
as they drool into the sand and water around me
drowns the ocean's crashings
until my inquiries create total sound

Nearby, a clockwork spider spins
in constant coils around tonight's dinner
and below me,
fish of varying color notice me,
nod politely, then continue
in their steady pursuits.

In streaks and splashes of metallic gleam
traffic slows and stops in rush hour's halt;
unsatisfied haste causes drivers to
fidget and sound their horns for private justice.
They seem born to a different Earth
and they continue to drive in waves
toward the thinning red horizon
until their solar chaperon will slip away
and by her absence, will open
the gate for ambitious sailors.

Within the hour, the sky and ocean
will meet and rise together in violet
to rule as one over a dreamscape
of progression.

So I pass time with the fishes
waiting to ask mother
about the process of atomics,
love,
and life after death.

An Uneventful Ride

Peter Gelman

With an illustration by Neal Skorpen

How slow I am, I thought leadenly as I felt my slow tires sink and sag in the gluttonous pavement. Heavily I climbed from downtown Portland up the hills to the west.

Preoccupied with this sense of slowness, and burdened by a sense of gravity's personal hatred for me, I missed the turn on 53rd to Forest Park and found myself on Cornell. I may have been distracted in my reluctant admiration of the rather remarkably square clouds. They seemed to form the shapes of each of the individual states of the USA, excluding Alaska and Hawaii.

Cornell is a narrow road that serpentines up through in the west hills of Portland. The two thin lanes pass lush green canyons of what seems to be rain forest (especially if it is raining) and passes through two narrow tunnels. On this occasion, that joyous time beloved by all cyclists and known as "rush hour" was beginning. Automobile traffic roared and screamed up and down the road in a most agreeable matter.

I noted with satisfaction that today, most of the cars sported the brand names of freshness, sunlight, and spirituality. (Yesterday's automobiles sported the themes of ripping, knifing, and mangling.) Today promised to be a pleasant, tranquil ride without any of the previous day's remarkable incidents.

Yet such was my optimism and pluck that I attempted to ride on a portion of the pavement that these cars required in order to bring their freshness, sunlight, and spirituality. All seemed well, for these were *enlightened* drivers, used to seeing cyclists on the road. They permitted me my one to ten inches of asphalt *without* hostility. The drivers did not honk as they brushed near me at enormous speed. They treated me with kind liberal smiles as they smashed through the atmosphere sending harsh pieces of it hurling against my flesh. While they had no objection if I wanted to ride in the mud, the ditches, or even the bits of rubble alongside the road, they did *not* insist on it.

It was I, the obnoxious bicyclist, who was antisocial and disagreeable. The drivers had admitted me into their club —their club being the whole road—due to a charitable outreach program. What was wrong with me? So what if I felt a constant dread of a fast heavy menace? Why not accept the smiles accompanied by blasts and blares? I must reluctantly confess that due to my unreasonable nature, I found the drivers' smiles unacceptable. I *tried* to think of kittens, and I *tried* to think of puppies, but it didn't work.

So I turned my bike off the road and stopped. The sneer on my face uncoiled, and my lips made a "fup-fup-fup" sound, the internationally recognized sound of resignation. I put my foot down on the ground, where it sank into deep gooey mud. I pulled it up again, and the mud hung on to my shoelace, undoing the knot.

When I pulled up my muddy shoe, I put down the other one, and then had to lift *it* from the mud. Of course, the mud dutifully pulled that knot loose, too. Who cares? I thought as I looked at my two muddy, unlaced shoes. I felt off my stride and defeated.

And now I must allow myself a personal and private observation; if you are squeamish and don't wish to see me standing before you bare, without all the false vestments of society, then look away! If you don't want to see me standing bare and posing in all sorts of contorted and yet attractive postures (I am speaking of my soul), please avert your eyes.

Being an extraordinary human being, and ever trying to reconcile myself to that difficult fact, I tend toward the perfectionism that is my true nature. Therefore I am easily frustrated, seldom meet my goals, train erratically and to no point. I am prone to give up my cycling trips easily, even at the slightest fall of a twig in my path. In fact, even a dandelion seed laying in my path is enough to put me in a horrible mood with scowls and reproachful glances at innocent passers-by.

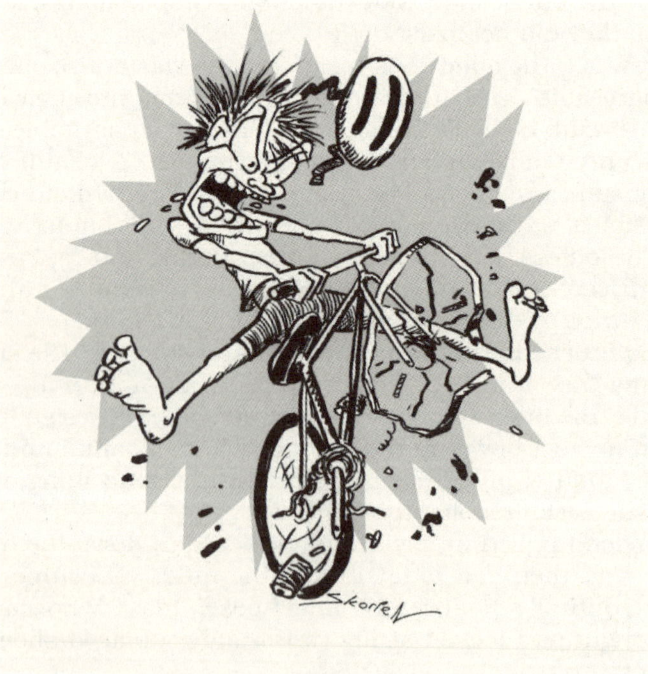

Worse, I hardly even know how to ride a bicycle sometimes; for instance, which way am I supposed to rotate the pedals? Why don't they stamp arrows on the metal indicating the proper direction?

Feeling disconsolate and ill-used as I stood in the mud of a Cornell Road ditch, I decided to accept my defeat and turn around.

No sooner had I crossed the road and started cycling again when I found my mood beginning to improve. A good mood, as well as velocity, is a necessary product of the geared motion of this machine, this peculiar machine that was once called the birotate chariot. Good Mood: That is what they should stamp on the gears of the bicycle, along with a suggestive directional arrow. As I birotated along, I suppressed a smile for upwardly creased cheek muscles are less aerodynamic than the sleek facial posture of frowning.

As I quickly picked up speed downhill I was already traveling ahead of myself, in fact I was home already. I was conducting mankind's old ritual of solace, pressing my lips to the noble grape, a grape gently but firmly crushed and fermented. I was already having the first phantom sip when an unusual event happened.

But before we get to the unusual event, let me tell you that I absolutely whipped down Cornell, peddling as fast as I could to keep up with traffic before I reached the tunnel. On my immediate left I could feel the perturbations of an automobile; a glance revealed that it was, of course, a white car (one named after a cloud with particularly gentle edges) with a Labrador Retriever poking its head out the rear window. On my right was a deep, narrow dirt trench trickling with a robust and nourishing trickle of water. Penned in, I needed to go faster. I engaged my full Nietzschean Überman determination: my feet whirled around in the pedals.

Suddenly—and this begins the unusual occurrence (as so often the word "suddenly" does) I felt a tugging at the top of my foot. The tugging increased. I glanced down and saw something that dismayed me deeply.

My shoelace had become caught between gear and chain. It was mashed and entangled in the teeth. My only option, given my obstinate and cowardly reluctance to hurl myself and my bike into the trees, was to go faster, hoping that the teeth of the gears would chew through the lace. However, the nylon shoelace merely wound around the gear. The gear gently but firmly pulled the shoe open as I began to cross a small stone bridge. The moment I crossed the bridge, and turned my attention back to the shoe, it popped off my foot and shot through space. Where did it go? I looked around frantically. I turned awkwardly in the saddle. It was nowhere.

With a tantrum of perturbed petrochemicals, the car-with-the-name-of-a-cloud sped around me and ahead. It disappeared behind a bend, leaving me to breathe its noxious fumes. As the car disappeared I thought I saw the silhouette of the dog gnawing on my shoe. I held my breath as long as I could, until the next unusual thing happened.

It didn't happen suddenly, I must admit. Time passed, as time tends to do. Over time—time, I mean, not in the geological sense of eons, which would be ridiculous, but in the human bike ride sense of minutes and seconds, I began to realize that I now felt a tugging at my other shoe.

Evidently my decision not to take the time to tie double knots after stepping in the mud was considered an act of hubris to the ancient gods that confound our pride. I couldn't believe it would happen again so I gritted my teeth and peddled on. My speed increased with my furious pedaling... The force pulling inward against my foot grew stronger with each revolution. Suddenly the shoe sucked off my foot, slipped around the teeth of my cog several times, disintegrating into cross-training dust and action-striped nothingness.

Barefoot I pedaled on manfully. In fact I even increased my speed. The road flew underneath me and all was wet and green around me, except for another car, a car with a name expressing spirituality, coming up behind me. I knew I had to increase my speed so that I would pass through the tunnel ahead of it. There wasn't room for a car in either lane plus me. I laughed with glee. I smiled and glared wildly. I cursed the ancient gods and dared them to do their worst.

Finally I entered the tunnel. The car-with-the-spiritual-name followed right behind me. I was quite nervous, and pedaled harder. At that moment I felt the horrible rumbling vibrations of a dump truck, no less, as it brutalized the atmosphere, coming full speed uphill in the other lane. The way was quite narrow but I successfully increased my speed. As the car-with-the-spiritual-name caught up to me so did the dump truck ahead of us, all timed to intersect together in the tunnel.

At that same moment, a sudden moment more sudden than it was a moment, I saw the gaping crater.

Potholes are, by and large, somewhat smaller than our State's glorious Crater Lake. Surprisingly, however, the smaller craters are a depraved tribe, while Crater Lake is sort of plucky and often cheerful. This particular Cornell Road pothole was hard and jagged and there was a hubcap lying beside it and—please note—a super-size fast food soda cup.

While some readers may argue that it would have been possible for me to dodge the pothole and the hubcap without weaving into the truck or slowing to be hit by the car-with-the-spiritual-name behind me, it would not have been possible for me to miss the hubcap and the pothole without also missing the opportunity to run over the super-size soda cup.

With the glee of Sherman setting fire to Atlanta, bringing liberation as well as destruction, not to mention resentful repercussions, I hit the wax paper cup; if you geometrists must know I neatly bisected it. As I cried out in triumph my tread smashed into the hubcap, knocking it aside as it plunged down into the pothole and smashed into the far edge. My front wheel crunched.

The exact aesthetic nature of that crunching sound I contemplated in the eternity of that small fraction of a second. As crunches go, it was not a good crunch. It was the kind of crunch one might call a worse than average crunch. But, in all honesty, it was probably not the worst crunch ever heard. For example, I imagine the sound the *Titanic* made when it hit that iceberg was one or two notches worse, possibly. The crunch of the mad swimming rabbit that bit Jimmy Carter's canoe might well have been worse. However, there was a certain subjective *quality* to my crunch.

One might even go so far into anthropomorphic interpretation as to say, it was a moderately *malevolent* crunch.

Enough with subjectivity! I knew instantly the exact nature of the damage. I knew that a third of my spokes looked like so much twisted lettuce shreds. The rim had caved in; the wheel now took the shape of a gibbous moon. As this type of moon is not often rolled in the mouth of poets, I had an idea that the front wheel would not roll for me either.

Hush! Let us not quibble over poetry! There was hardly any time for me to consider the questionable merits of the gibbous moon shape, for at that instant, at my velocity (perhaps thirty-four miles an hour) there was only one thing to do, and I had to do it immediately!

There was no time to digress in idle contemplation of imponderable mysteries! For in that same fraction of a moment, as my hapless bike lurched toward disaster, the dump truck rumbled by, its slipstream blasting me and slapping my face deliberately. Yet in the moment of my great disaster I remained balanced, determined and calm.

With my massively powerful proletarian forearms I heaved up on the handlebars, throwing my weight backward and driving the tender soles of my bare feet harder than ever onto the pedals. My rear wheel pushed forward faster than the rest of the bicycle, forcing the broken front wheel up into the air.

Never was a wheelie more triumphant, more demonstrative of the indomitable human spirit, than my wheelie.

As I emerged from the tunnel on my wheelie, I took a quiet moment to observe nature. For I feel, as an urban man who more often rides his bicycle between automobiles than Druids, that I cannot afford—whatever the danger—to neglect the concentrated absorption of the natural joys. I don't mind risking my reputation and employment by admitting that I like pine trees and wild flowers, green vistas and even the occasional flapping bird. Neither am I adverse to bees buzzing. I hope you won't think ill of me for saying so, for I have noticed that some of you tend to sneer at "window obstacles" which you call trees and you flash your middle finger at Bigfoot and tell pointless knock-knock jokes to spotted fawns. I know, it's our Constitutional right as Americans to hate and destroy Nature in all its forms. I however will risk your outrage by admitting that I am inclined to quietly appreciate the pointiness of the three or four mountains we can see from Portland, depending on which mountains were in the neighborhood that day to get their points sharpened.

I noted, while riding my triumphant wheelie, that the clouds had utterly disappeared. It had turned into an exquisitely beautiful day. It was perhaps the first day of spring. Birds flitted among the trees. Sometimes the birds flitted and sometimes they flapped; there were two kinds of birds out that day. Two kinds! I stared, amazed. And suddenly, more inspired by the flapping than the flitting, I had a thought.

As down, down the hill I rushed, I thought about that thought. As thoughts go, it was surprisingly snappy. It was peppy, too. It contained neither legal warnings nor technical references. (Down, down, I descended on my wheelie through sloping neighborhoods.) The thought even avoided coarse language and pop culture references. (Down, down I flew, toward the zone of boutiques and coffee franchise.) These are but a few of the characteristics that do not describe my thought. Noting these noncharacteristics while looking at the doctors and lawyers in their leather jackets ambling around their motorcycles outside the coffee franchise (as I approached a streetcar track), I congratulated myself on my thought. As I considered the judiciousness of those congratulations—neither too scanty nor lacking in zeal—I realized I had forgotten the thought.

Just then, with a horrifying "ker-plunk", the streetcar track knocked the lone tire under me. My wheelie tottered! The manly grip of my buttocks upon the saddle struggled—weakened—and slipped! My feet slid backward—until only the toes remained, clutching the pedals! And as is so often the case, some of my toes weren't really trying very hard. As just a few of the community leaders among my toes held my wheelie upright, my buttocks floated in weightlessness. I saw the motorcycle gang of doctors and lawyers gasp with horror and sympathy. I, too, also filled with horror and sympathy, also gasped! Heroically, my bare toes crawled up the pedals. Thus my feet found the pedals again. Jostled, I knew I had forgotten my thought, but still I was preoccupied with restoring the Copernican equilibrium of my buttocks upon my wheelie. A moment later, I forgot even that I had forgotten the thought, both the snappy and the peppy parts; and somewhere in his dark castle, Oblivion laughed.

Rolling along, a moment passed in quiet ignorance. I wheelied along in a state of perfect balance with the benevolent universe. I felt the urge to whistle a few lines of an old sea chantey. But for some reason, I did not. A moment later, I heard a remarkably snappy and peppy voice—my own—uttering these memorable words:

"*What a snappy little planet for my peppy little bike!*"

I laughed a mad happy laugh and increased my speed even more. I was, in fact, making very good time.

Near Death

Janice Levy

I almost got killed today. A man mowed down a one way street, aimed his car at mine, and stopped within an inch of plowing into me. On purpose. I've never seen him before and probably never will again, except in my dreams tonight. My windows were closed, the radio was on. I don't know what he sounds like or what language he speaks. But, tonight he'll hear my voice. I'll be in a tank, the Batmobile, possibly a tank. Will I brake? Maybe. I'll have to sleep on it.

I'm afraid of heights and too-open spaces. I'm anxious about flying; what if I just had to open a window or if the plane began to shrink? Perhaps I've seen too many movies. Yet, when I sit in the bleachers of Yankee Stadium or use my frequent flyer mileage, rational thought takes over. What are the odds, I think. So although I still fear that if there is a disaster, it'll strike me, I know there's an excellent chance I'll make it to extra innings.

Random violence is something else. I can't prepare mentally or physically. There are no definite whys or wheres. The what-ifs are too great. My friend's t-shirt reads, "when I have time, I'm going to have a nervous breakdown." In a bizarre way, she says, that keeps her going. On the other hand, random violence doesn't give you time to luxuriate in its sheer terror.

When the man screeched away, I thought of the police. What could I say? His car was red. His eyes were blue. His teeth were yellow. "A primary experience," I actually said out loud at the time. Then I thanked God. I figured He had more in store for me, I hadn't seen all the colors of the rainbow just yet.

My immediate relief was that I was alone. "He doesn't know who we are," I would have repeatedly reassured my daughter. "He doesn't know where we live." My husband and son would have morphed into Gibson/Glover and an ensuing Lethal Weapon car chase. And if my dog had been injured, my bite would have been worse than her bark.

Do I hate this man? I won't give him that power. I'd rather view him as a snapped bungee cord. Being a writer by trade, I've concocted plot lines and genres; an X-file escapee, a sour milk romance, a showdown in the old corral? Give me time and I'll twist it into a funny picture book. Really.

Am I angry? He switched t-shirts on me, raised the ante, tilted the odds. Now when I see a red car, I'll flinch. Not enough to get off the road. But just enough to move into the right lane.

As I lay in bed, I take calming breaths. Rationally, I know I'll get over this. It's random, inconsequential. But I don't turn off the light. I won't. I can't. Not yet.

A Spin of the Wheel

Nicholas Wees

Roger stands on the grass between Delta street and the park and watches as his older brother steps out into the crosswalk.

They've come to the park from their house two blocks away, looking forward to a few care-free hours, excited because of the uncertainty that comes with the absence of parents and of structure, but also secure because of the familiarity of park and brother and an unconscious faith that all things are safely ordered. But Roger is too young to go alone.

Of course, he prefers going with his mother, prefers the certainty that she carries in her worn shoulder bag—her "park bag", she calls it—the way she explains everything, strings events together into a coherent pattern; but when he hears his brother ask for permission to go to the park, Roger knows that permission will only be granted if Nathan agrees to bring him along, and, with or without his mother, he always looks forward to the sounds and space of the park. On the way there he runs ahead, yelling.

They always take the same route to the park and always look both ways twice before crossing Delta street.

But when they reach the other side Roger looks back and sees Pitter, their cat, on the far sidewalk. A quiet thrill crackles through him. "It's Pitter," he says. She has followed them—come to play, he thinks, away from parents; maybe she will finally speak to him, as he suspects she can.

Nathan, however, isn't so excited. Old enough to understand the dangers of the road, though he knows that Pitter often wanders far from their house on her secret feline meanderings, he is shocked to see her there, so out of place by the busy street. He hesitates, unsure.

Roger takes a step towards the road but Nathan grabs hold of him. "Wait here," he says, "I'm gonna go get her. Just wait here." He turns and walks quickly away from Roger. When he reaches the crosswalk he looks both ways and sees that the traffic is clear. Just then Pitter starts towards him, stepping gingerly off the curb.

"Stay there, Pitter. Stay." Nathan has his hands out, motioning to the cat to stop. As they approach each other in the middle of the street the cat lies down and looks up at the boy.

From where he stands, his small shoulders at attention, his brown eyes alert, Roger sees his brother reach down and scoop up the cat. Then there is a sound.

Nathan is saying something to the cat but Roger can't hear him; he looks up, the sound growing, rumbling over him, and sees the driver's eyes. A moment before they were gazing lazily towards the park; now, directed straight ahead, they snap into focus. Roger tries to call out; his mouth opens and closes but makes no sound. He can see the cat's eyes flash amber as she leaps from his brother's arms; he can see the driver's eyes, wide and white, but from here he can't see his brother's bright green eyes.

The sound roars over him, grinding, hurling itself into a screeching fury. Then there is a dull thump.

* * * * *

By the time Roger made it out of the Energy building—well past five o'clock, as usual—the rain had slowed to a steady cold drizzle. He climbed into his car and let the engine warm up for a minute before pulling out of the parking lot.

As the car hissed through the growing dusk, he settled in

to the drive. Twice a day he made the forty minute commute between his house and his office in the Energy building. It was only at this time that he was truly alone and he savored every moment of it.

Throughout the rest of the day, he was in constant demand—by his wife, his daughters, his co-workers; everybody needed something from him, he who was always on time, forever ready, perpetually competent and reliable. But now, as daylight dimmed, he could lapse into a contemplative state of mind, allow his thoughts to roam; the cares of the day lost their immediacy and drifted into perspective; and though his mind floated freely as he drove, at all times he was physically alert and fully in control of his vehicle. Every reflex was poised, ready for action.

He approached driving the same way he approached everything: he learned its rules so intimately that it became second nature to him. This allowed him to look ahead, to foresee when and where someone or something might deviate from the rules of the expected. More than inconvenience, whether minor irritants or major hardships, Roger was committed to warding off the unexpected. As long as he knew what was coming, he was in control.

As he cruised along the highway, his mind would slip away; his thoughts would drift apart, fragmenting into a million shards of consciousness, all peering out into the unknown, familiar scenes replaying themselves with infinite variations. While his nerves and muscles continued to function in an automatic and informed way, what had been Roger dissolved into possibility.

And then one tiny spark of consciousness would burst out, growing brighter, expanding, eclipsing all else and a single thought would coalesce, gain materiality, and Roger would come rushing back into being. These thoughts could be of anything, from the mundane to the sublime.

The first thought that materialized in such a way that evening was of his daughter, Kate. She was to take her driving test that day after school. He was amazed that she was old enough to drive and it made him feel sad in a strange way. But he knew she would pass the test without any problems; he had taught her himself.

"The most important thing that you have to remember," he'd told her during that first lesson, "is to always assume that someone is going to make a mistake." This was not so much a lesson about driving as about life in general.

"If you're always aware of other people's potential screw-ups then you can avoid them." It was not enough to know the rules and to follow them; one must be able to anticipate when they would be broken, when the certainties of life would collapse, and so, be able to thwart the unforeseen, to banish the uncontrollable.

Though he did feel a certain anguish at the thought of Kate driving, he was nonetheless proud that she could be entrusted with an automobile.

And though he didn't love cars, he'd long accepted their inevitability. He never did any kind of work on his own car; he left that to his trusted mechanic. Roger never changed the oil or put air in the tires; he never checked the timing or installed a new air filter, though he could have. He preferred to pay someone else to carry out those duties, but he had, in fact, an extensive knowledge of auto mechanics. As with all things that he was dependent upon, he learned everything he could about them. Whether it was his car, his computer, or the planning of a cruise vacation, he kept himself informed on every detail. He would not be caught off guard.

The rain had all but completely stopped as he pulled off the expressway. Only a grey formless mist clung to the trees. There were fewer houses, some small farms, and here and there, roads leading up to new subdivisions. He stopped at a convenience store to pick up ice cream and cat food.

Back in his car, he let his mind drift back into the fuzzy warmth of semi-consciousness, knowing that it would come to an end when he pulled into his driveway fifteen minutes later.

* * * * *

Warm air fills the car, pouring out of vents on the dash and floor. But for it, the windows would be cloaked in thick droplets of moisture.

Henderson street—its long winding curves ease along the golf course. A lazy asphalt river, skirting tree-lined properties, its tributaries with names like Princeton Court and Emerald Crescent. The fan in the car blows, soothes. Its warm electric hum advances and recedes, blends with the drone of the motor, the rush of tires on wet pavement.

Rockheights—a pause at the four-way intersection, a left hand turn, and the engine rumbles back up into gear.

The car hisses on as trees, houses and fields blur past, though muscle and nerve are tensed, alert to any disturbance in the steady hum that radiates from the car out, rippling through the fog-laden night. There is no center to the ever expanding sound. Particles gaze back in at themselves, from across eternity, and then, suddenly, come rushing, hurling back together, in an instant harmonizing into a single note, a single thought.

Headlights sweep across the glistening road, and there, in the shadows, ahead, a sudden burst of movement brings eye to attention, rouses feet and hands. Sound telescopes in on itself. The hum is a roar. The roar is a scream.

All shreds of consciousness have imploded into reaction. A small grey form materializes, green eyes flash in the headlights; the roar hiccups and screeches to a halt.

* * * * *

Roger sat for a moment, his hands still tight on the steering wheel. As soon as he stepped out of the car the dampness enveloped him. He walked around behind the car; the air was cool and still.

Then he saw the cat lying by the ditch. He walked over and crouched down beside it; it was moving, but just barely. Its hind legs were completely crushed; its eyelids drooped and its mouth slowly opened and closed. Roger slipped his jacket off and wrapped it around the cat as he carefully lifted it into his arms.

The closest house was fifty or sixty yards back down the road. As he walked towards it, he was vaguely aware of a bell sounding, trying desperately to remind him that he'd left the door open and the engine running, but it was drowned out by the silence that was now washing over him.

When he reached the driveway, the cat made a quiet noise, a soft hissing plea. Roger lifted the jacket and looked at the bleeding furry wreak. Its small body shook slightly and its jaw moved again, but no sound came. Suddenly, he was aware of how light it was.

It should have known, he thought. Living on a semi-rural road, where cars are fewer but faster; it should have known to beware those roaring creatures that hurled down the road with fearful regularity. It should have known to look both ways.

But it could not have known. Of course, Roger knew this—that the cat could never understand the rules of life, could not have predicted its chances on that road; he also knew that he had slipped up, that his reflexes had not been fast enough. He could have avoided it, if he'd been watching more closely. But why? Why couldn't the cat have seen the danger? He had tried his best—even the safest projections might be proved wrong sometimes.

It should have known. Nathan did; he looked both ways. Then he knelt down to pick her up. It should have known to not trust the street, to not relinquish command of its own small destiny and put itself in the path of ultimately unpredictable—as though with a will of their own—mechanical forces. Nathan knew. He abided by the rules. Only it wasn't enough to rely on the rules, the trick was to anticipate when they would fail. Nathan knew; but he was betrayed when he knelt down to pick up the cat. She had leapt clear. Roger could see her eyes as she turned to jump. He could see the man's eyes. It should have known. Now it was a small bundle of *was* in his arms.

Roger mounted the steps, cradled the cat in one arm and knocked. Just as he was about to knock again, a light came on behind the door. It was opened by an old man in a plaid house coat and worn slippers.

The man looked at him then at the cat. Before Roger could speak, the man reached out his hands. Roger wanted to tell him how it had happened, how careful he always was, but the old man said nothing and only took the cat from him, holding it close to his chest.

"I'm so sorry," Roger began, but the sound didn't quite match the meaning. The old man only looked at him for a moment, expressionless, then turned and closed the door behind him, leaving Roger alone on the porch.

The damp air was growing colder and from deep inside, he felt a tremendous sorrow, a bewildering agony welling up. Then it burst in a rushing torrent, flooding over him, cascading, spilling out into the night.

Traffic Songs

Lee Henderson

A Mighty Fortress

Tune: *Luther's Hymn: Ein Fest Burg*
Key: G

A mighty fortress is my car,
My great strong box of steel,
My shiny sporty four-by-four,
My urban sex-appeal.
This safe on wheels may crash
And small cars I may smash,
Some slobs may be dispatched
But I shall not be scratched,
My sport ute' is my fortress

The world can see how cool I am,
How excellent is my lifestyle.
All tongues proclaim the brands I buy
My choices are such high style.
Now get out of my way,
Bike burns in spandex gay
And stupid old ladies!
And leave the road to me
In my shiny fossil fortress.

Car Insurance Rates

Tune: *Old Gray Mare*
Key: D

Car insurance rates are unfair to me.
I don't drive much you see,
At least not habitually.
Car insurance rates are too much for me;
I don't drive much at all (3 times)

My workmate Hal
drives a great big SUV.
He drives it incessantly
from here to eternity.
My workmate Hal
drive each day from Langalee
And pays the same rates as me.
He pays the same rates as me (2 times)

Guys like Hal endanger society,
Roarin' them SUV's,
Makin' the children flee,
Guys like Hal
are spewin' that smoke ya see,
And they should pay twice the fee.
But they pay the same rates as me.
I'm giving them a subsidy.

Most every place I go ain't very far;
I don't hardly use my car,
My bike can get me thar.
Most every day I grab the handlebars;
I wanna pay by the mile,
I think I could save a pile,
If I could pay by the mile.

Car insurance rates are too much for me;
I don't drive much at all (3 times)

Traffic Jam

Tune: *Workin On the Railroad*
Key: C

I'm just waitin' in the traffic
Half of my workin' day,
So I had to get a cell phone
To take some calls along the way.
Had to move out to the suburbs,
Had to buy an SUV,
Now I'm waitin' in the traffic
For half an eternity.

I'm steppin' on the gas,
I'm leanin' on the horn,
Ragin' in the traffic jam, I am.
I'm standin' on the brake,
I'm flippin' you the bird,
I'm ragin' in the traffic jam.

Oh. How I hate this drivin'!
Gotta get in to work (it's so far)
Oh, how I hate what I do there,
But I gotta pay for my car.

God damn traffic jam!
Ragin' in this jam I am.
God damn traffic jam;
This is where I am.

Traffic Lullaby

Key: A
Tune: *Brahm's Lullaby*

Please watch out! Hang up the phone!
Please drive slowly and carefully.
Kids here walk to school,
Parents bike to work,
Streets are narrow, you can see.
Cool the road rage, lay off the gas,
You don't need to go so fast.
Cool the road rage, lay off the gas,
You don't need to go so fast.

Traffic calming is so good
For a cool urban neighbourhood.
People walk and play
On the streets each day,
and life is looking good.
Cool the road rage, lay off the gas;
No one needs to drive through fast.
Cool the road rage, lay off the gas;
No one needs to drive through fast.

The traffic circle looks so sweet
In the place where our two streets meet.
Let's plant flowers and make it ours
With tiles of all the children's feet.
Cool the road rage, lay off the gas;
No one needs to drive through fast.
Cool the road rage, lay off the gas;
No one needs to drive through fast.

The Tragic Tale of Steven and His Spam Tin-Colored Fox on Wheels

Stephanie Scarborough

It had been many a night that Steven had salivated, ogled, and wept like a sissy over her pin-up that was tacked ever so carefully to his equally carefully spackled wall. The only flaw was that pesky, scantily-clad supermodel who was obscuring his view of the front right fender. The curvaceous (dare he say—voluptuous?) body, the tinted windows, and the gold paint that shone like a radiantly shellacked Spam tin in the sunlight. Sparing all frivolous words, it made Steve pleasantly happy.

And now, after working three jobs, investing in several high-interest CDs, switching to store-brand Cheez Doodles, and borrowing twenty dollars from his friend, Jack, Steven was about to own the supreme culmination of automotive evolution. Just the mere thought of being able to caress her velvety red interior—day in and day out—at his whim made him, to be as sparing as possible with words, quite pleasantly happy.

He had camped out at the dealership to ensure his punctuality, his hoard of money in his vintage lunch box and a plastic valve in his heart (too many darn Spamburgers). At eight A. M. Steven was banging on the front door. At 8:01, the police were dragging him off the car lot. At 9:27, Steven was back, prepared to wheel and deal or beg and grovel, which ever method worked. The dealer was friendly —disturbed by Steven's compulsive behavior and foul body odor—but friendly because he could see the large bills peeking out from under the lid of Steve's lunch box. Two hours and seven minutes later Steven's lunch box was considerably lighter and the dealer's pockets equally heavier, but more importantly, Steven was sliding into the car. And the front right fender was all he'd imagined and more. The suppleness of the padded steering wheel between his hands made him . . . . . . pleasantly happy.

He let the velvety red seat conform to every curve and crevice of his derriere, sat back and just sighed; he had eaten ten complimentary doughnuts in the show room. Jelly filled, no less. And Steven sat there in the car, experiencing his own private little ecstasy, until he was urged by the dealer to drive it off the lot because it had closed over an hour ago, and everyone really wanted to go home.

Drive the car? Why, sticking the key, which was the same radiant, shellacked Spam tin color as the car, in the ignition hadn't even occurred to him. Would it be too much to handle? Maybe he should have started doing this around noon.

So he removed the curiously smooth key from his shirt pocket and gingerly slid it into the ignition as though he was a gourmet chef slicing into an octopus-tofu pâtè that was about to be served to a mother-in-law whom he feared very much. He sighed again, and not because he had eaten ten complimentary jelly doughnuts that morning. The dealer was growing weary. Steven then was, much to the weary dealer's surprise, quite fastidious with turning the key, and the dealer joyously bade him adieu.

On the road, Steven was equally fastidious with the care in which he handled his shellacked Spam tin-colored beauty.

Never too much gas at once; never too much brake at once. Every sign, light, and speed limit was religiously obeyed. Steven spent much of his time feeling quite pleasantly happy.

But after about five minutes on the road, Steven found there was little time to wallow in this pleasantly happy feeling for there were few other drivers who were as fastidiously fastidious as he was about obeying such frivolous guidelines as speed limits. Perhaps they thought all those signs and stop lights were simply suggestions from the Department of Public Safety and could be taken into consideration at one's whim. And a few gave Steven's automotive masterpiece the respect it deserved. For instance, in the Pastry Mart parking lot some punk who couldn't regulate how far the door on his junkmobile swung open dinged the side of Steve's immaculate pride and joy, leaving a mark that resembled a Cabbage Patch Kid face, and didn't even bother to apologize. Steven would have broken down and cried except there were people around and quite a few at that. So he climbed ever so carefully back into his car, set his box of crème horns in the passenger seat, and left, feeling somewhat agitated. On the highway, a sadistic-looking lady in a mini-van honked at him and suggested he either "move it or lose it, bucko." Steven could not understand why she was so disgruntled. The speed limit was seventy miles per hour, and according to his glow-in-the-dark speedometer he was going exactly seventy. Her speed, however, undoubtedly surpassed eighty, perhaps even ninety, yet he, being the easy-going person he was, did not feel the need to honk frantically and holler, "Excuse me, ma'am! May I suggest you decelerate just a tad, perhaps? You're exceeding the posted speed limit by twenty miles!" But he did find himself gloating a bit when the police pulled her over. He felt compelled to stick his thumbs in his ears and chant, "Nanny-nanny-poo-poo!", but refrained from doing so. Feeling a bit better, Steven was able to relax and enjoy the remainder of his drive home, though he did have one last stop to make. He needed to stop by the Spatula Barn for some last minute gift ideas. Mother was wanting the Henry VIII Collection which included six spatulas with detachable heads, while kiddies enjoyed the Day Glo spatulas. He even signed his father and himself up for the Spatula of the Month Club.

Steven felt much (though not completely) better after his shopping spree. The only thing that made him happier than having a spatula in his hand was the breath-taking sight of his spanking new, almost flawless . . .

. . . Steven dropped his armload of spatulas and felt the need to go change his boxer shorts. A large, hairy man with poor personal hygiene habits in a once-white tank top stood by the remains of Steve's almost flawless automotive masterpiece, gnawing on a crème horn. "This your car?" he asked. "Hope you got insurance. I was backin' up the big rig, and, silly me, I wasn't expectin' your car to be parked in the darn fire lane an' I just plowed right into her. I helped myself to the crème horns. Hope you don't mind. The flies were startin' to gather, and I couldn't let good food go to waste. You know—"

Before the big hairy man with the poor hygiene habits could finish, Steven, a mere dust mite in comparison, had leapt and latched onto the large man like a rabid Chihuahua. And he had little intention of letting go. He began futily beating his comparably puny fists against the truck driver, being more of a nuisance than a threat. Nevertheless, his persistence and the disturbing fervor with which he pounded against the truck driver prompted the manager of the Spatula Barn to call the police. Steven had to be tranquilized before they could remove him from the truck driver, and he now resides in the local equivalent of a "funny farm" where he sits in his padded room and stares at his pin-up with that radiant car that once was his, mumbling incoherently about "no insurance" and "that darn fire lane" and other such matters.

Roadkill Bill

Ken Avidor

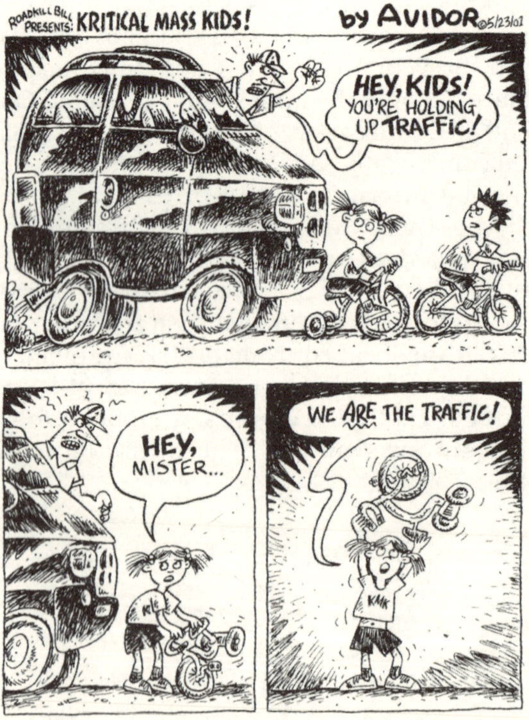

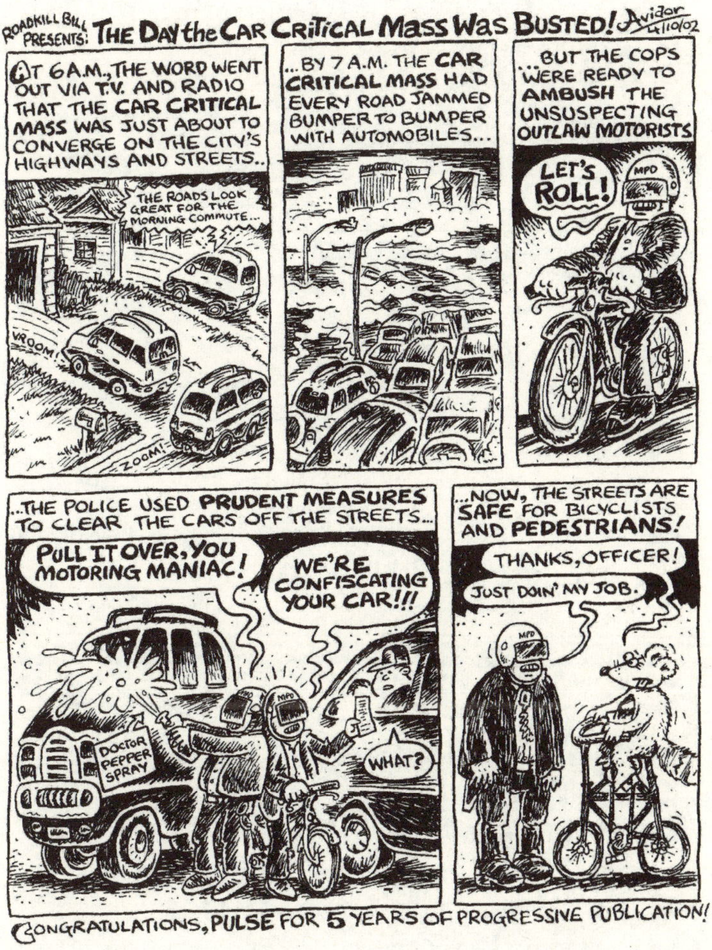
ROADKILL BILL PRESENTS: THE DAY the CAR CRITICAL MASS WAS BUSTED!
Avidor 4/10/02
AT 6 A.M., THE WORD WENT OUT VIA T.V. AND RADIO THAT THE CAR CRITICAL MASS WAS JUST ABOUT TO CONVERGE ON THE CITY'S HIGHWAYS AND STREETS..
THE ROADS LOOK GREAT FOR THE MORNING COMMUTE...
VROOM!
ZOOM!
...BY 7 A.M. THE CAR CRITICAL MASS HAD EVERY ROAD JAMMED BUMPER TO BUMPER WITH AUTOMOBILES...
...BUT THE COPS WERE READY TO AMBUSH THE UNSUSPECTING OUTLAW MOTORISTS
LET'S ROLL!
MPD
...THE POLICE USED PRUDENT MEASURES TO CLEAR THE CARS OFF THE STREETS...
PULL IT OVER, YOU MOTORING MANIAC!
WE'RE CONFISCATING YOUR CAR!!!
DOCTOR PEPPER SPRAY
WHAT?
...NOW, THE STREETS ARE SAFE FOR BICYCLISTS AND PEDESTRIANS!
THANKS, OFFICER!
JUST DOIN' MY JOB.
CONGRATULATIONS, PULSE FOR 5 YEARS OF PROGRESSIVE PUBLICATION!

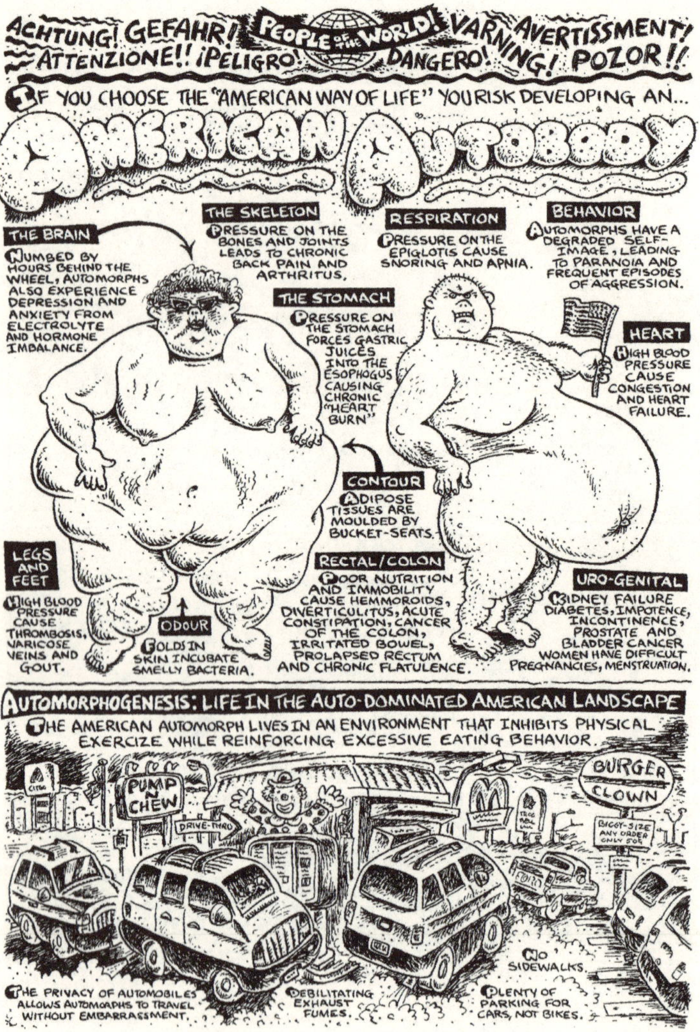
ACHTUNG! GEFAHR! PEOPLE OF THE WORLD! VARNING! AVERTISSMENT!
ATTENZIONE!! ¡PELIGRO! DANGERO! POZOR!!
IF YOU CHOOSE THE "AMERICAN WAY OF LIFE" YOU RISK DEVELOPING AN...
AMERICAN AUTOBODY
THE BRAIN
NUMBED BY HOURS BEHIND THE WHEEL, AUTOMORPHS ALSO EXPERIENCE DEPRESSION AND ANXIETY FROM ELECTROLYTE AND HORMONE IMBALANCE.
THE SKELETON
PRESSURE ON THE BONES AND JOINTS LEADS TO CHRONIC BACK PAIN AND ARTHRITUS.
RESPIRATION
PRESSURE ON THE EPIGLOTIS CAUSE SNORING AND APNIA.
BEHAVIOR
AUTOMORPHS HAVE A DEGRADED SELF-IMAGE, LEADING TO PARANOIA AND FREQUENT EPISODES OF AGGRESSION.
THE STOMACH
PRESSURE ON THE STOMACH FORCES GASTRIC JUICES INTO THE ESOPHOGUS CAUSING CHRONIC "HEART BURN"
HEART
HIGH BLOOD PRESSURE CAUSE CONGESTION AND HEART FAILURE.
CONTOUR
ADIPOSE TISSUES ARE MOULDED BY BUCKET-SEATS.
LEGS AND FEET
HIGH BLOOD PRESSURE CAUSE THROMBOSIS, VARICOSE VEINS AND GOUT.
ODOUR
FOLDS IN SKIN INCUBATE SMELLY BACTERIA.
RECTAL/COLON
POOR NUTRITION AND IMMOBILITY CAUSE HEMMOROIDS, DIVERTICULITUS, ACUTE CONSTIPATION, CANCER OF THE COLON, IRRITATED BOWEL, PROLAPSED RECTUM AND CHRONIC FLATULENCE.
URO-GENITAL
KIDNEY FAILURE DIABETES, IMPOTENCE, INCONTINENCE, PROSTATE AND BLADDER CANCER WOMEN HAVE DIFFICULT PREGNANCIES, MENSTRUATION.
AUTOMORPHOGENESIS: LIFE IN THE AUTO-DOMINATED AMERICAN LANDSCAPE
THE AMERICAN AUTOMORPH LIVES IN AN ENVIRONMENT THAT INHIBITS PHYSICAL EXERCIZE WHILE REINFORCING EXCESSIVE EATING BEHAVIOR.
PUMP N CHEW
DRIVE-THRU
BURGER CLOWN
BIGGY-SIZE ANY ORDER ONLY 50¢
NO SIDEWALKS.
THE PRIVACY OF AUTOMOBILES ALLOWS AUTOMORPHS TO TRAVEL WITHOUT EMBARRASSMENT.
DEBILITATING EXHAUST FUMES.
PLENTY OF PARKING FOR CARS, NOT BIKES.

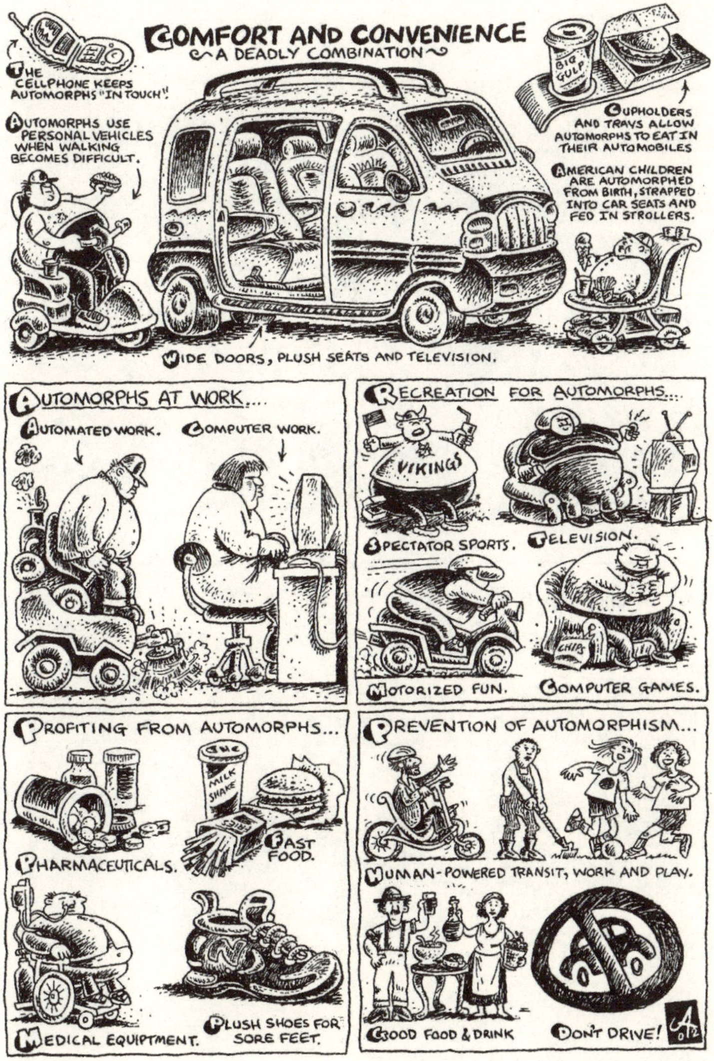
COMFORT AND CONVENIENCE
A DEADLY COMBINATION
THE CELLPHONE KEEPS AUTOMORPHS "IN TOUCH"!
AUTOMORPHS USE PERSONAL VEHICLES WHEN WALKING BECOMES DIFFICULT.
CUPHOLDERS AND TRAYS ALLOW AUTOMORPHS TO EAT IN THEIR AUTOMOBILES
BIG GULP
AMERICAN CHILDREN ARE AUTOMORPHED FROM BIRTH, STRAPPED INTO CAR SEATS AND FED IN STROLLERS.
WIDE DOORS, PLUSH SEATS AND TELEVISION.
AUTOMORPHS AT WORK....
AUTOMATED WORK.
COMPUTER WORK.
RECREATION FOR AUTOMORPHS...
VIKINGS
SPECTATOR SPORTS.
TELEVISION.
MOTORIZED FUN.
CHIPS
COMPUTER GAMES.
PROFITING FROM AUTOMORPHS...
MILK SHAKE
PHARMACEUTICALS.
FAST FOOD.
MEDICAL EQUIPTMENT.
PLUSH SHOES FOR SORE FEET.
PREVENTION OF AUTOMORPHISM...
HUMAN-POWERED TRANSIT, WORK AND PLAY.
GOOD FOOD & DRINK
DON'T DRIVE!

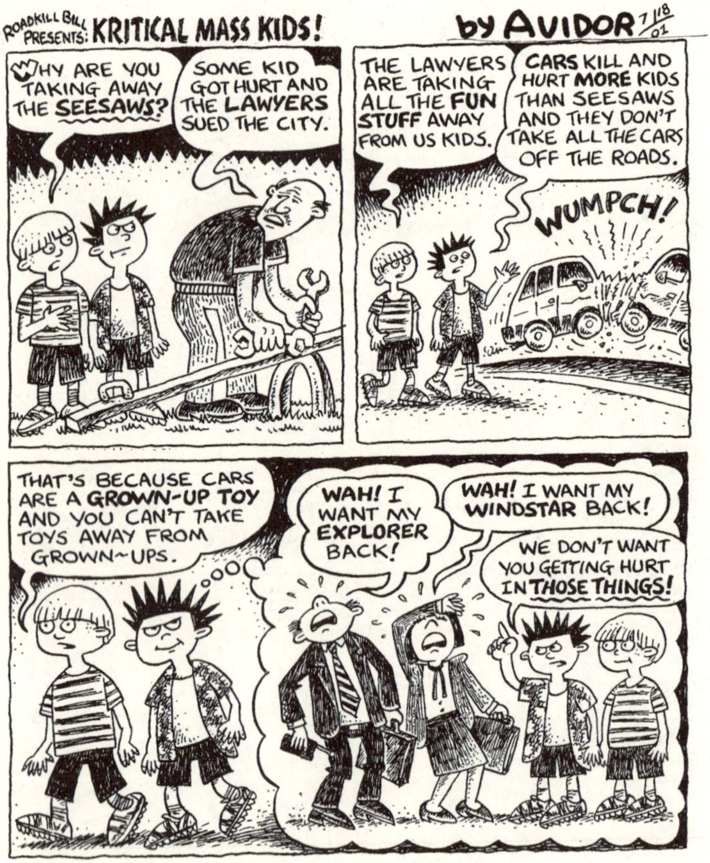
ROADKILL BILL PRESENTS: KRITICAL MASS KIDS!
by AVIDOR 7/18/01
WHY ARE YOU TAKING AWAY THE SEESAWS?
SOME KID GOT HURT AND THE LAWYERS SUED THE CITY.
THE LAWYERS ARE TAKING ALL THE FUN STUFF AWAY FROM US KIDS.
CARS KILL AND HURT MORE KIDS THAN SEESAWS AND THEY DON'T TAKE ALL THE CARS OFF THE ROADS.
WUMPCH!
THAT'S BECAUSE CARS ARE A GROWN-UP TOY AND YOU CAN'T TAKE TOYS AWAY FROM GROWN~UPS.
WAH! I WANT MY EXPLORER BACK!
WAH! I WANT MY WINDSTAR BACK!
WE DON'T WANT YOU GETTING HURT IN THOSE THINGS!

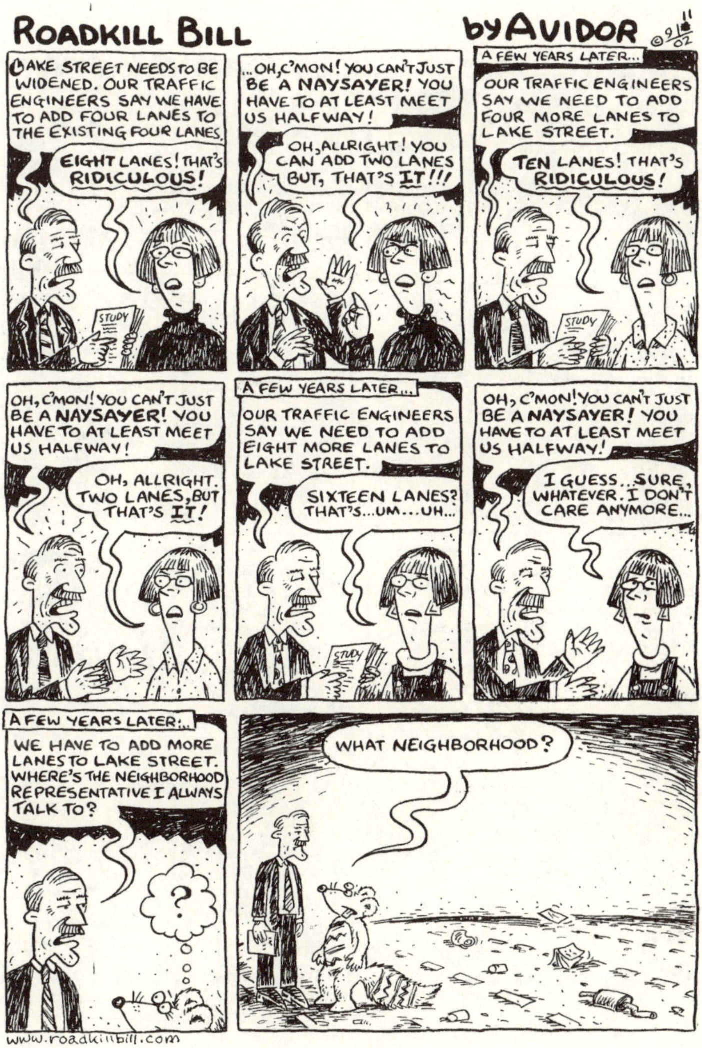
ROADKILL BILL
by AVIDOR ©9/11/02
LAKE STREET NEEDS TO BE WIDENED. OUR TRAFFIC ENGINEERS SAY WE HAVE TO ADD FOUR LANES TO THE EXISTING FOUR LANES.
EIGHT LANES! THAT'S RIDICULOUS!
STUDY
...OH, C'MON! YOU CAN'T JUST BE A NAYSAYER! YOU HAVE TO AT LEAST MEET US HALFWAY!
OH, ALLRIGHT! YOU CAN ADD TWO LANES BUT, THAT'S IT!!!
A FEW YEARS LATER...
OUR TRAFFIC ENGINEERS SAY WE NEED TO ADD FOUR MORE LANES TO LAKE STREET.
TEN LANES! THAT'S RIDICULOUS!
STUDY
OH, C'MON! YOU CAN'T JUST BE A NAYSAYER! YOU HAVE TO AT LEAST MEET US HALFWAY!
OH, ALLRIGHT. TWO LANES, BUT THAT'S IT!
A FEW YEARS LATER...
OUR TRAFFIC ENGINEERS SAY WE NEED TO ADD EIGHT MORE LANES TO LAKE STREET.
SIXTEEN LANES? THAT'S...UM...UH...
STUDY
OH, C'MON! YOU CAN'T JUST BE A NAYSAYER! YOU HAVE TO AT LEAST MEET US HALFWAY!
I GUESS...SURE, WHATEVER. I DON'T CARE ANYMORE...
A FEW YEARS LATER...
WE HAVE TO ADD MORE LANES TO LAKE STREET. WHERE'S THE NEIGHBORHOOD REPRESENTATIVE I ALWAYS TALK TO?
?
WHAT NEIGHBORHOOD?
www.roadkillbill.com

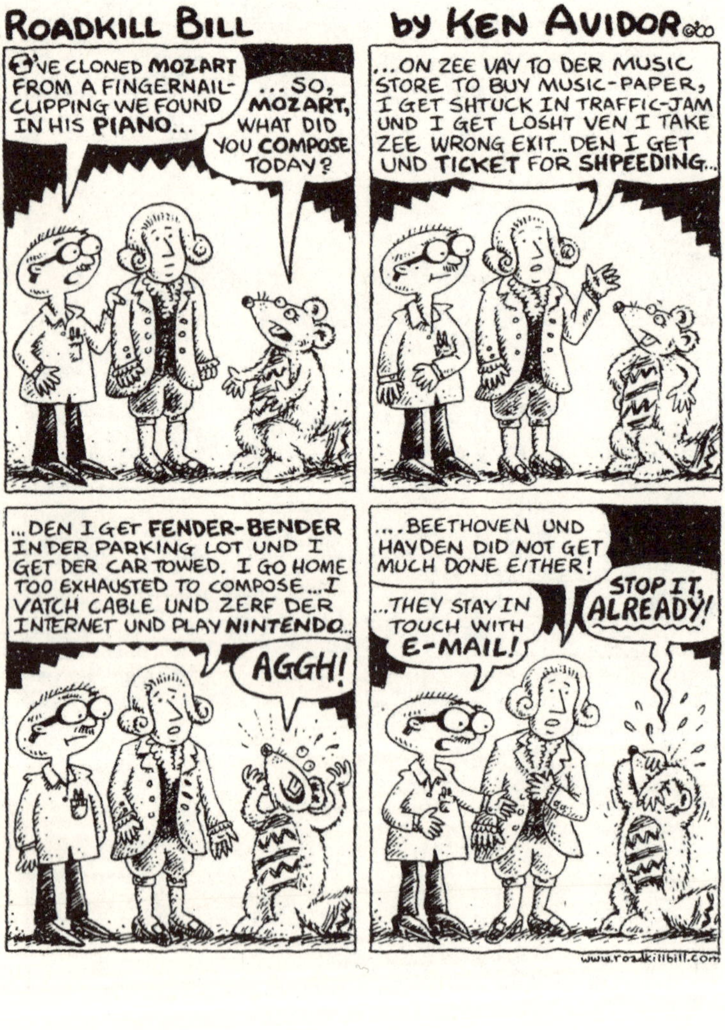
ROADKILL BILL
by KEN AVIDOR
I'VE CLONED MOZART FROM A FINGERNAIL-CLIPPING WE FOUND IN HIS PIANO...
...SO, MOZART, WHAT DID YOU COMPOSE TODAY?
...ON ZEE VAY TO DER MUSIC STORE TO BUY MUSIC-PAPER, I GET SHTUCK IN TRAFFIC-JAM UND I GET LOSHT VEN I TAKE ZEE WRONG EXIT... DEN I GET UND TICKET FOR SHPEEDING...
...DEN I GET FENDER-BENDER IN DER PARKING LOT UND I GET DER CAR TOWED. I GO HOME TOO EXHAUSTED TO COMPOSE... I VATCH CABLE UND ZERF DER INTERNET UND PLAY NINTENDO...
AGGH!
....BEETHOVEN UND HAYDEN DID NOT GET MUCH DONE EITHER!
...THEY STAY IN TOUCH WITH E-MAIL!
STOP IT, ALREADY!
www.roadkillbill.com

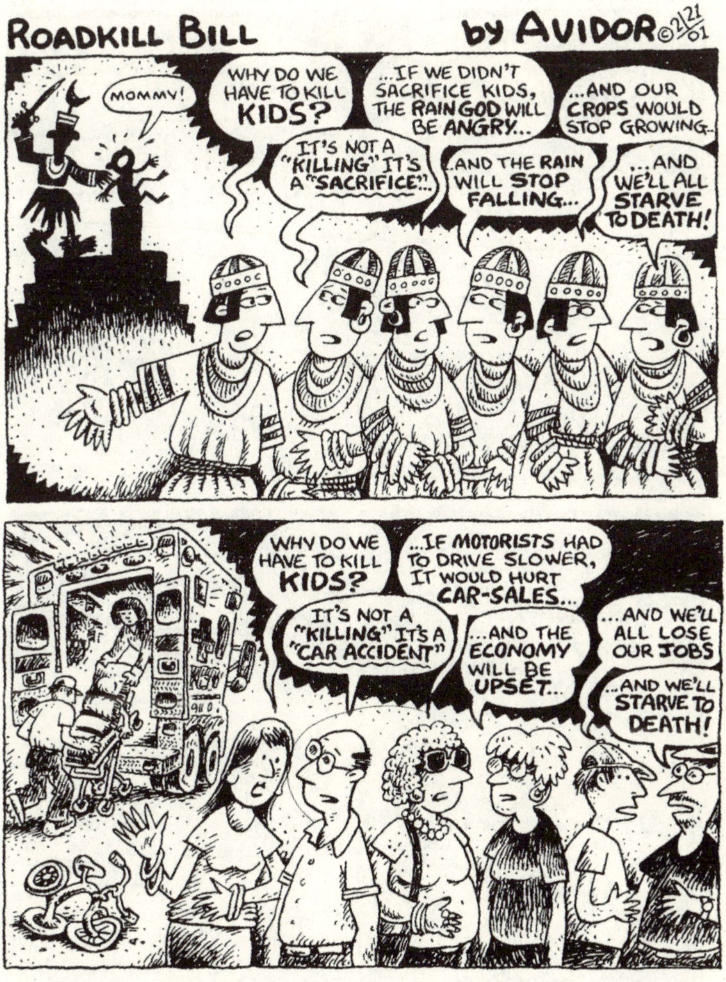
ROADKILL BILL
by AVIDOR ©2/21/01
MOMMY!
WHY DO WE HAVE TO KILL KIDS?
IT'S NOT A "KILLING" IT'S A "SACRIFICE"...
...IF WE DIDN'T SACRIFICE KIDS, THE RAIN GOD WILL BE ANGRY...
..AND THE RAIN WILL STOP FALLING...
...AND OUR CROPS WOULD STOP GROWING..
...AND WE'LL ALL STARVE TO DEATH!
WHY DO WE HAVE TO KILL KIDS?
IT'S NOT A "KILLING" IT'S A "CAR ACCIDENT"
...IF MOTORISTS HAD TO DRIVE SLOWER, IT WOULD HURT CAR-SALES...
...AND THE ECONOMY WILL BE UPSET...
...AND WE'LL ALL LOSE OUR JOBS
...AND WE'LL STARVE TO DEATH!

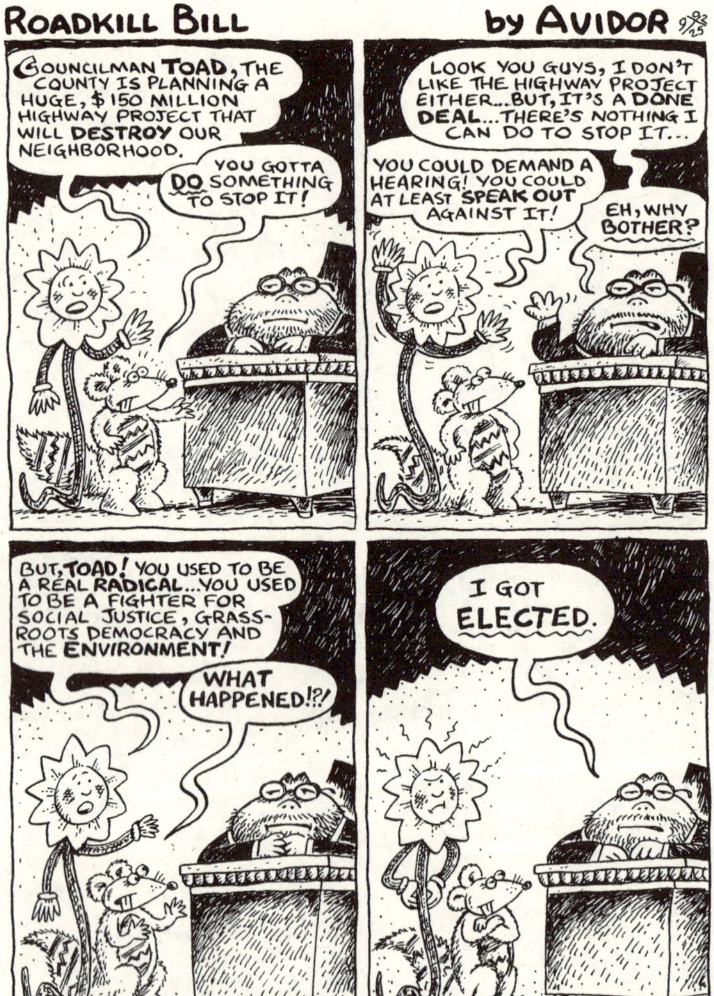
ROADKILL BILL
by AVIDOR
COUNCILMAN TOAD, THE COUNTY IS PLANNING A HUGE, $150 MILLION HIGHWAY PROJECT THAT WILL DESTROY OUR NEIGHBORHOOD.
YOU GOTTA DO SOMETHING TO STOP IT!
LOOK YOU GUYS, I DON'T LIKE THE HIGHWAY PROJECT EITHER...BUT, IT'S A DONE DEAL...THERE'S NOTHING I CAN DO TO STOP IT...
YOU COULD DEMAND A HEARING! YOU COULD AT LEAST SPEAK OUT AGAINST IT!
EH, WHY BOTHER?
BUT, TOAD! YOU USED TO BE A REAL RADICAL...YOU USED TO BE A FIGHTER FOR SOCIAL JUSTICE, GRASS-ROOTS DEMOCRACY AND THE ENVIRONMENT!
WHAT HAPPENED!?!
I GOT ELECTED.
STOP THE 35W ACCESS PROJECT!
WWW.ROADKILLBILL.COM

Crady's Second Ride

Greg Blanchette

Crady hit the jive point six ticks early—buggy, 'cause a ride never showed too soon. So he tucked behind the plastic hedge in front of the Tikki Hut, in case of passing cops, and straddled his bike to wait. Better early than late, after yesterday's fiasco. Blue'd called last night, said don't worry, *bay-bee*, happens to everybody, but even at phone resolution Crady could see the blue eyes laughing. On his first ride! His face flushed hot when he imagined the look Honch would give him when he hooked today.

Crady remembered the first time he ever saw a ride. He'd wrapped his refresher courses back east and moved out here on a whim. It was where most of the softshops were, so finding the job had been a one-bit shift. He'd been standing near the front of a packed stretch bus jerking along in heavy traffic, and some kids in the back started barking: *A ride! Cut 'em off! Squish 'em!*

He'd craned his neck to glimpse two hunched, purple-clad bodies sweeping past the bus. Out front they merged with a pair from the other side, legs pumping, and flowed like purple mercury into single file and squirted up the gap between the lanes. *Wow!*

"Suicidal fools," the old codger jammed against him had muttered.

"Why do they all wear purple?" a girl sitting by his knees asked her companion.

Crady knew; he had an old bootleg ROM of *2 Wheels Good* at home. "Psychological armor," he'd blurted out in response. "Driver sees a big, solid block of color, he instinctively avoids it."

The girl'd looked at him incredulously, the animation in her flat oriental face dwindling to a point and vanishing, like an old-style TV switching off. She huddled with her friend and giggled. Too bad, Crady thought, she was kind of cute.

Now a guy in a dirty overcoat walked out of the Tikki Hut, slurping coffee, and ogled Crady. Crady stared back till he passed the hedge, then looked over the traffic, now sitting inert as a volcano about to blow, and checked the time on his cycle computer. Still minutes to go.

That bus ride had been months ago, and a long dull stretch it would've been, no doubt, but for one thing: Among his 200-odd fellow cyber-sweats, one had been rumoured to ride. Crady found this "Blue," made a friend, and began filling his empty evenings and off days at the velodrome: bike handling, speedwork, impact. Blue'd flashed him the current version of *2 Wheels Good*, drilled him on signals and procedure. Crady even moved hopefully to East Vee, where Blue and the others lived. A long shot, but what did he care where he lived?

Eventually he met the rest of the ride. Dice, rangy and taciturn, characteristically cool. He worked an overwater bike shop that catered to gadget freaks—hydraulic transmissions, electronic suspension, nav units—though his own rake was all business, nothing fancy, custom-built by himself for riding rush.

Honch (she pronounced it "Honk") had turned her short beefy frame toward him, shaken her spike-blonde hair, and taken in Crady's thin legs with contempt. Blue, his freckled face split by a grin, told him later she had twice turned down national-team scouts. "Legs like pit bulls," he said. "Coulda gone pro—sponsorship, Olympics, the works, but she likes working the gym and riding rush. We put up with the attitude."

Beaneater, the usual fourth, was at the moment reclining in a hospital bed—which happenstance, along with Blue's urgings, led to Crady's *first* ride. Which had rolled smooth as a square wheel, Crady thought, now grimacing, peering over the hedge. Cars, cars everywhere, parked on lawns, idling at the light, lurching ahead in spastic squirts of smoke glass and plastic chrome. Cars packed so tight it seemed there was no room for anything else, till you looked closer and saw the gaps between bumpers, the half-metre between lanes—the unclaimed flux of space in which rush-riders ran their routine.

Now the computer beeped, and Crady snugged his charcoal mask and set his shades. After a few seconds, Blue's minor-third yodel cut weakly through the hiss of high-pressure radials and the bleep of horns. Crady clipped his cleats in and shot down the sidewalk, glancing over his shoulder. Honch at point, she and Blue crossed Nanaimo St. in a deuce, taking up the same space a small car would. Crady synch'ed speed, hopped off the curb, and slid in next to Blue. An irate blast of horn from the car behind made Crady glance back in annoyance, then check himself: *Don't acknowledge, don't provoke.*

"Not bad today," Blue yelled over the rumble, click, and hum of a hundred different engines. Crady couldn't tell whether he was joking or not. Vehicles of all sizes, bumper-to-bumper across five lanes in both directions as far as you could see. He said nothing, concentrating on that burn in the knees as he warmed back up. Another good reason not to be early.

Honch hadn't even glanced his way. They cruised along dead even with traffic. Dice hooked at his jive spot just past Victoria Drive, swapped to point, and as a foursquare brick they rolled on down First Ave.

Crady was looking over at Dice, riveted by his fluid cadence, the way his head swiveled back and forth like a scanning radar turret, when suddenly there was a tremendous jolt on his front wheel. Pothole! Crady slewed over and slammed into Blue, who neatly absorbed the collision.

The blare of horns was instantaneous and disorienting.

Crady's kevlar thigh plate had taken the impact with Blue, and all those painful hours of impact practice at the 'drome kept him rubber down. He was a little dazed, and a lot more alert, as they caught the light at Commercial Dr. and crested the downhill, now rolling easy with traffic. Groenik!

Crady had ground glass in his gut as they closed on Clark Dr., the intersection where he'd lost it yesterday. Blue at point then, pressing hard for the amber. They'd almost made it too: last vehicles through as the light flopped red, except a trot in a crossing lane had nosed out early. Dice and Blue ducked, Crady and Honch couldn't make it. The sibilant skid of her hard narrow tire and Honch's bellow were the sounds Crady remembered, though he couldn't possibly have heard them in the howl of horns as he bounced his front wheel off the trot's door. But Honch—Honch had rewritten the laws of physics so far as Crady'd known them: She'd skidded around parallel to the car, leaned out to plant a hand on its hood, and without a hitch in the line of motion she *vaulted*, bike and all, right over its front end. She landed on her wheels and, without so much as a backward glance, stood up in a smoking sprint, reeling in Dice and Blue ahead. That was the rule, Crady knew—*no waiting, ever*—but it sure felt harsh and lonely at the moment.

He'd just stood there, one foot on the pavement, and it'd been a toss-up which was more astounded, him or the trot. Then the light flopped green and three lanes of traffic both ways sprang into motion, Crady right smack in the middle. He'd been clipped twice before he got himself into an alley to catch his breath and curse. Hopeless by then to chase the others. No choice but to snake it home by alley and sidewalk and catch a bus to work, where he arrived fifty minutes late. He'd punched extra hours till he was hoarse to make it up, and it had taken Blue near half an hour on the vid last night to talk him into another try.

But today at Clark there was no ork with an itch. The light flopped amber as they approached the intersection and Dice, thinking maybe of Crady, called for brakes. Signaling like a train crossing, they came to a stop right at the head of the line. Some jack behind leaned on his horn, which blared out a crass bugle call over and over, and rolled down his window to yell.

Earplugs muted the noise, but it was still an uncomfortable position—wide open to the view of any renta-cops in the cross traffic, though it was unlikely they'd stop for small-time fines like a ride. Of more concern was the jack, working himself into overdrive because they sat there so casual in front of him on "his" pavement. When the light flopped he'd be rabid for all that open road (maybe half a block to the back of the next queue), and even a sprint wouldn't be fast enough for him. To peel over single file by the curb was asking to get door-handled as the cars squeezed by. Only thing was to brick the lane and hammer, trusting to autonomic conditioning to keep the driver from ramming them up the rear.

According to *2WG*, that was the one thing you *could* trust about driving: all instinct, very little reason involved. Which is why people found it soothing under relaxed conditions, and maddening in the high-pressure city commute. Sure, Crady knew, they were people just like him in control of these cars, but in the interests of survival they were best thought of as pea-brained reptiles. But reptiles, at least, who heeded certain primitive laws, such as: Don't attack what's bigger than yourself; Keep the herd packed tight, but not too tight; and above all, Preserve thy paint-work.

Industry pundits had long predicted that the introduction of low-emission engines would revolutionize urban transportation, and for once they were right. What changed was the iota of guilt people once felt at over-using their cars. It vanished. Soon after the auto-makers began pumping out alt-fuels, hybrids, and full electros, more traffic hit the streets: first a little, then a lot; then solid streams of cars, jamming every available space. People got stressed, irrational, mean. Crady'd seen an experiment on the net: You take a big cage, put say 20 rats in it; they lead happy lives. Then you put 40 rats in, with more food and water: crowded, but life is still okay. At 60 it's chaos, but you can see that some kind of order, of rat civilization, still prevails.

Suddenly at 80—catastrophe. They go vicious, insane, attacking each other, eating their young....

Dice glanced back, finger-signing "When it flops we go," and Crady nodded automatically. They were already moving as cross-traffic screeched to a stop, and it was all he could do to hang with the brick. Jack, despite the histrionics, was slow on the gas and they opened a gap, then slid single-file to blast non-stop all the way down Terminal Ave. between creeping lanes of traffic. The corner at the Skytrain station was gridlocked as usual, choked with vehicles trying to turn or waiting for pedestrians to cross. Some stared, a few jeered, as the ride wormed between lanes to a staccato of horns from incensed drivers. One threw open his door to block their passage and stepped out, beer-belly belligerent. But the sight of four fit and possibly desperate outlaws in masks and mirror shades made him think better of a heroic stand, and he dove hastily back inside, double-locking his door.

Traffic lurched forward again a block north on Main St. Dice signaled "lane change left," and Crady thought, *long way today, floatbridge must be patrolled.* Riding a brick, they caught the advance arrow and took the turn onto the Viaduct. Dice cooked it up the ramp to the elevated roadway, speeding up as they slid into file.

This was The Stretch, a kilometre-long pipeline into the heart of downtown—close as it comes to open road in the city. For the drivers it was *ladies and gentlemen, start your engines,* and they made the most of it, getting up over 80 klicks some mornings. The stink on a windless day was overpowering. The industrial filter masks Crady and the others wore could barely handle the ozone from the electros, and nearly half the buggies were still internal-combustion, converted to alt-fuels from methanol to horse piss. Now with even the cheapest electro costing over 50K, people showed surprising ingenuity at keeping their old IC heaps running. Propane was the worst. Crady could not figure how something that stunk so bad could be touted as e-friendly.

Traffic was thick and fast this morning. Dice didn't even

try for a brick, just hammered along the edge of the right lane which (a rarity) was wide enough to share. Crady clicked into gear, tucked, and cranked.

The pink light of the rising sun behind him glinted off the windows of the bank towers ahead. The same light washed the fabric bubble roof of the stadium, a patched, grubby, sunburnt belly. Crady's gaze floated down Honch's ass, her pumping thighs, her calves tight in electric-blue lycra.

Dice drifted back alongside, millimetres away, blowing hard. The angry rip of a nervous driver's horn startled Crady. Blue gasped out a yodel as he took his turn up front. Crady felt the burn spread from butt to thighs.

The driver alongside was waving. Crady glanced over at him. The man gave him a thumbs-up, grinning, then suddenly twitched his joystick and Crady swerved involuntarily away. The other riders held position, Honch extending a fist and rapping a warning on the guy's fender. The plates on the edge of their gloves only made noise, but the spikes over the knuckles would gouge paint if the ork persisted. Crady looked angrily over as the bald little bugger pulled ahead, laughing uproariously.

He looked a lot like Crady's rich Uncle Skeet, who had once cornered him, years ago, like in a scene from some bad movie. "I only got one word of advice for you, Charles," he'd said, "A-I." Then proceeded to elaborate drunkenly on that "word" for an hour. This was in the early days of Groenik's Hypothesis, when the sky suddenly seemed the limit in artificial intelligence and a smart young knowledge engineer might go anywhere. Few'd been smart enough to foresee that AI would bootstrap itself up so quickly and convincingly that soon even the *halfway* interesting work would be done by the machines themselves, requiring only three overlapping shifts of smooth-talking loopers to keep them in touch with market reality. Some didn't mind the softshops—Blue, for instance. But he was new to it, give him another couple years.

The wail of tires from close behind made Crady curse his wandering attention, and he braced for a smash of metal and glass. Dice hollered, a note of urgency in his voice.

What'd he say? Sounded like "needle-neck," meaning what? Crady glanced back and spotted the problem at once: Two cars back and gaining, a big old turd-brown IC boat, blotchy paint, ominous South Surrey "rebel" license plate on the dash. Hunter: Guy who cared nothing for his body-work or paint job, who lived for the havoc he could create on the roads. Usually you just turned off, or pulled over and stopped for half a minute. But stuck on the Viaduct between traffic and a concrete crash fence? Crady wasn't sure how nervous he should be.

Honch was yelling at Blue. He dropped back, and his glance at Crady as he drifted to the back of the paceline transmitted tension better than words could. Crady was suddenly scared, and he needed all the adrenaline he could juice to hang with Honch, who was out of the saddle and churning. They were running dead even with traffic, Needle-neck just a car back now, pounding his horn gleefully. But could she hold that pace another half-klick? Crady wasn't sure he could; he was maxxed already, fighting the urge to rip away his mask for air. He zoned in on Honch's tire, centimetres ahead of his. He didn't even realize the predica-ment until they were in it.

The Viaduct spit traffic into the downtown core up a gen-tle but significant grade. With traffic moving fast, not even Honch at full song could hang with it here. The crash fence came to an end, but the sidewalk was busy—another Sky-train stop—and to leap the curb at speed meant wipeout, possible injury. There was no cross-street to turn onto. And Needleneck was gaining. Just one option remained: change lanes toward the middle of the road, out of reach of the hunter's car.

Honch stabbed out a signal and peeled across the nar-row gap between two cars. Blue next, then Dice, and Crady weaseled across when he could, wincing at the blast of horn and the scant clearance. Dice and Blue ducked again, putting a lane between them and Needleneck.

Honch was blocked. They were slowing now with the hill and the maneuvering. Crady cranked till it felt like his knees would blow. He sensed the battered front of Needle-neck's car easing up dead close on his right—the tremor of an unbalanced tire, a pounded-out dent in the fender. Trapped!

Honch suddenly ducked right, into the impossible space ahead of Needleneck's car, taking her chances with the sidewalk rather than traffic. A big chrome bumper bullet nipped her rear tire as she swept across, and time for Crady did one of those slo-mo deals. He watched her fight it, countersteering into a sideways drift, and then go down in a tangle of legs and handlebars. Needleneck slammed on the brakes and as Crady shot past he saw Honch go half under the front of the car!

He had no idea how he did it, in the confusion of horns and the smash of at least two rear-enders somewhere behind, but he suddenly found himself facing back up the road beside the stopped car, glaring in at Needleneck. The boat was so old it had non-tinted windows, and Crady saw clearly the lean face, short carved-out hair, the protruding Adam's apple. Surprise lit Needleneck's face, along with a hair-trigger mix of worry and delight, as he gaped back and forth from the front of his car, where Honch could not be seen, to Crady straddling his bike.

Honch was screaming now, the lower registers of shock climbing rapidly to pain. Crady looked Needleneck in the eye and jerked his thumb. "Back up!" he hollered. Needleneck stared at him, mouth open. The cacophony from the line of stopped cars behind grew to impossible volume, although traffic next lane over hadn't even slowed down.

Crady repeated his gesture. Needleneck suddenly lunged sideways on the big bench seat, stretching out to fumble with the glove box. *Uh-oh*, Crady thought, and half-turned to flee. *No waiting, ever!* Needleneck's foot slipped off the brake. The old boat had an automatic tranny, still in "Drive", and it began to inch forward up the incline. Honch's screaming grew to hysterical pitch.

Crady grabbed the door handle—locked, of course. Without conscious thought he drew back his fist and punched, whole body behind it, *through* the barrier like he'd seen years before in a karate vid. The glass spiderwebbed but held. Another punch and it disintegrated, showering the inside with granules. Crady reached past the steering wheel and fumbled for the shift lever. He saw Needleneck's white face glancing back, wide-eyed now as he dropped a small aerosol can to the tattered carpeting. Crady got two fingers on the lever, slammed it into "Park", then back one notch to "Reverse". The boat shuddered and backed slowly down the hill. Crady raked his arm on a jagged edge getting it out the window.

Honch's scream was agony now as the car backed off her, and Crady suddenly hoped he hadn't done something stupid. The boat crunched lightly into the car behind as Crady hobbled to Honch.

She was already on her feet, cursing. Her helmet was impressively shredded down one side. A hamburger rash speckled the lycra over one shoulder. The kevlar plate had saved her thigh, but the calf showed blood. She was hopping on that leg and favouring the other.

"You all right?" he shouted in her face, though she clearly wasn't. A knot of pedestrians on the sidewalk watched, raptly.

She growled in rage and pain, plucking gingerly at the short dark brand of a hot pipe across her calf.

"Can you ride?" Crady yelled. "We've got to get out of here!"

Honch turned in Needleneck's direction. "You pencil-sucker! I'll snap your—"

Crady glanced back, remembering the aerosol can, probably Mace or pepper. The driver of the car behind was standing at the shattered window screaming at Needleneck.

Crady pulled Honch's bike upright. It seemed to be okay, except the handlebars were twisted off-center. He jammed her front wheel between a leg and the frame of his bike and wrenched the bars more or less straight. He picked up her shades from the road. The gawkers on the sidewalk were yelling, edging onto the pavement.

A sudden scream from behind, the sharp stink of pepper. Needleneck'd let the other driver have it full in the face, and the guy staggered backwards, hands on his eyes, straight into the line of moving traffic. A blast of horn, a sick crunch, and he spun to the pavement by Crady's feet, writhing. Holy sweet macro mother!

Crady grabbed Honch's arm and yanked her around. Tears streamed down her white face now, her lips twisted in pain. But the eyes were lucid, focused, angry—maybe, he thought, a hint of a plea for help. He handed her the shades, and with shaking hands she put them on and hitched up her mask. "Get on!" he yelled, shoving the bike at her.

She responded with a bellow of rage, and swung a leg over.

He clipped in and rode. Easy at first, up along the curb till they crested the hill, glancing back frequently. She was doing all right so he sped up—the cars charging into the empty lane were going by too fast and close for safety.

Safety, he thought, ridiculous word under the circumstances. He glanced back again. Honch was right behind him, emptying her water bottle on her burnt leg. He wondered whether the lycra was fused to her flesh, and the thought dropped a lurch in his stomach.

He sped up more, suddenly realizing the position he was in: rush, the gut of the city, a rank novice with one injured second, little idea where to go or how to get there. Pull off and hide, maybe, some alley or doorway. But there were cars and people everywhere. With the tights, the bikes, they'd stick out like a nine-bit byte. And with that circus behind, the cops'd be looking ... there'd be explanations, charges, seizure of the bikes. Boot it, the only option was to push on.

Honch shouted something, her voice lost in traffic noise. He glanced back. She pointed ahead, and when he looked, a lucky break: a few blocks ahead, the bright blue blotches of Dice and Blue sitting tall at a stop light, looking back. They wouldn't wait, of course, but maybe he could keep them in sight and just follow. He wondered how much Honch could sustain.

She shouted again, and when he looked, she gave the sign for "deuce." Of course—traffic was thick, no sense creeping down the gutter. The on-screen theory of *2WG* was beginning to make sense. Checking over his shoulder, he stuck his arm out in a tentative signal, answered by a thunderous blast of horn. A low red Italian electric brushed close, so close he had to tap it off. "Tentative" was clearly useless on these streets, and at the next scant opening he signalled once and moved in boldly. Honch followed, taking up her diagonal so they filled the lane. Crady ignored the horn behind, the threatening snarl of a racing IC engine. Groenik, he thought, you could *die* out here if you didn't know what you were doing.

That next light—was it Richards St. or Granville? And did it just go yellow, or was it about to flop red? Now he knew why Dice's eyes never stopped roving. The car ahead rushed the light, then braked hard at the last moment and Crady had to stop fast. Honch slid up beside him and they leaned on each other's bikes.

"You okay?" he yelled at her, looking into her shades and wishing he could stop asking that. A bit of blood streaked her left arm.

She looked at him sharply as if to say, "Guess, moron," and reached down to pop her front brake loose—wobble in the wheel, probably. "Straight on past Burrard," she shouted back, so obviously she meant for him to lead. He nodded.

The light flopped and the car ahead burned some rubber. Crady held his acceleration down, despite the roars of disapproval from behind. He squeaked through the Howe St. light, and then caught red at the next, while Dice and Blue crossed Burrard and swung left out of sight.

Traffic was incredibly dense. You could've walked from curb to curb on plastic and glass, except for the incessant jockeying. On the bus he paid scant attention, but out here unprotected it was bare-ass frightening. The two of them didn't look like a whole lot of mass and, like gas expanding into a vacuum, vehicles continually tried to edge in upon them. It took a lot of nerve, and some stern tapping, to hold them at bay, all while keeping his eyes peeled for turn signals, road cracks, traffic lights, pedestrians... He glanced fondly at the buses they passed.

Fortunately the pace was undemanding, except for the

jackrabbit starts. They held a deuce to Burrard, where half the traffic turned off for a floatbridge on-ramp, and then slid into single file as they pushed on straight. The road turned steeply up to the left and he came up fast to the top of the short hill. A cement truck sat there, waiting for a break in traffic.

Crady braked hard and Honch bumped him from behind. "Sorry," he signed over the rumble of the truck's exhaust, "no viz."

She replied with another vicious look and a shout, of which he caught "... signal, jerk?"

Gigahertz bitch, he muttered, not without a twinge of admiration.

The truck pulled out in an eruption of diesel smoke and Crady trailed it, signing pointedly for a deuce in its wake. Fast-moving here, down a four-block stretch that always seemed to be ripped up, cranes and construction trailers filling the curb lane. Scouting ahead, he recognized a new problem: the approach to the Stanley Park causeway. Four lanes narrowed through a blind ess into three, then two. It was a textbook meatgrinder. Better to just pull off into the park itself and take the footpaths, he was thinking in dismay, tortuous and shameful but it wouldn't get anybody killed, when suddenly there was a flash of color and Dice and Blue slid in alongside. Of course—they'd ducked in front of one of the parked trucks and waited! Dice took point, Crady fell back beside Honch, they were a brick again, and a cold jet of relief shot up his chest. Blue, eyes bright as LEDs over his pushed-down shades, winked at Crady and yowled. He was grateful for the infusion of cheer.

They hauled brazenly down the anachronistic HOV lane, sandwiched between buses and vans, scoping the traffic in the other lanes. Here the approaching drivers jostled, frowning more intently as they braced for the battle of the merge.

As their lane tapered to an end Dice signaled "can-opener." He chose a bumper gap and slid his rake into it with nothing to spare. Blue followed suit, then Honch, then Crady, till they rode four abreast like sardines in a sandwich. Crady and Blue began to wobble, holding the puzzled eyes of the woman in the trailing car, and then fell back in unison, prizing open the gap until a short brick fit. The woman submitted in good grace, but the ork behind her went apoplectic at the intrusion. Crady grinned to himself, ignoring the horn with blatant magnanimity. The can-opener was his favourite maneuver, as close as riders got to imposing their will on traffic.

They made it through the pot-holed ess, rapped off two pushers at the second merge, and held the brick onto the causeway itself, which cut a disreputable swath through Stanley Park en route to the bridge. The ragged forest each side held in the fumes and noise like whorehouse curtains, and it was all grind here: jarring pavement, a steady uphill steeper than it looked, the serious glute-burn that feels like driving nails into your butt. No fun at all, and then Honch sheered away, yanked up her mask and puked her breakfast all over the sidewalk. Crady held out his water bottle, which she accepted with a nod, and took her empty one to shove in his bottle cage. They crested the hill and rolled easier down to the Chang Shen Motor Corp. auto-toll bridge, now rebuilt and enlarged for the third time, its cracked and convoluted "bike lane" still so useless for a fast commute that they stuck to the pavement.

The hard bit was the long, uphill half of the bridge, where a gap invariably opened in front of the straining ride. The space as invariably closed again on the downhill section, but trailing drivers never seemed to make the mathematical leap that told them they'd lost no ground. Instead they went brain-stem nuclear at the sight of open road, and only taps, spikes, and cold nerve kept them at bay.

Once off the bridge, though, tension blabbered out like a released balloon. The road surface improved and the bulk of the crazies were left behind in the city core, or locked in the bridge line-up trying to get there. They rode fast for another few klicks, hard work, but after city centre it was honest release—a few speed bumps to hop, a belligerent fuel cell Euro import to be stood off, a signaling panel truck to be let warily in.

At Pemberton Ave. there was a miraculous combination of good viz and no bus at the stop, and they peeled round the turn in formation like a squadron of fighter jets. One tricky left onto a side road choked with double-parked cars, and then 'Dullah, smiling, waved them around the gate and up the corkscrew ramp of the parking high-rise in the vicinity of Wheel-Tek Cycle, where Dice worked; the ⟨Real⟩ World Gym, Honch's haunt; and A**Z, the softshop that ate Blue and Crady's days.

Crady hung back while Dice and Blue bracketed Honch, scoping her injuries. Then they were on floor six, end of the ride, stuffing their bikes in the lockers. Winding down, Crady again felt that thrill he first felt that time on the stretchbus. It was just the commute to work, for Groenik's sake, supposedly the dullest part of an monotonous day. But today was a whole new subroutine. Today he *rode*—muscle, wit, and a nine-kilo rake versus rolling tyrannosaurs with brains the size of walnuts. And he done good. Standing there, he still felt his heart pounding. These days, you only ever got your heart rate up at the company aerobics session, or the movies, or during sex—and who had sex anymore? Not him, anyway.

But then Honch, her face paler than usual, looked at him as the door slid open, nodded, wagged his water bottle in the air, and disappeared into the elevator, and a new thrill seized him. Was that the trace of a smile on her face?

On the walk to the stairs, Dice said, "Honch's out a few days till the leg heals, but if you want to ride a triple in the meantime ...?" Crady tried not to grin too obscenely and said yeah, sure, no prob. And Blue, tap-dancing in his cleats as they started down the stairs. *Bay-bee!* he shouted, grabbing Crady by the helmet and shaking.

The Autoholics Anonymous Car-Free Pledge

Carbusters

"Hi, my name is Jane and I'm an autoholic." Admittedly, the idea of a 12-step self-help program for car addicts might seem amusing at first, but this is intended to be totally serious. That is, we're quite serious about our aims but we're light-hearted enough to call ourselves "recovering car addicts." As we all know, many serious social and environmental problems can be linked to car use. We address this by being proactive and working to change our individual and collective behaviour.

The Twelve Steps of Autoholics Anonymous

1. We admitted we were victims of the transportation industry seduced by speed, status, and sense of power—that we had become isolated and passive consumers of distances whose shape and length we could no longer control. We admitted we were not free but enslaved

and powerless in this situation.

2. We came to believe that Human Power and Access By Proximity could restore a balance in ourselves, and between the human and the natural environment.

3. We made a decision to build up our will to create personal and social change, and to transform our lives.

4. We made a searching and fearless inventory of the impacts of our car addiction.

5. We admitted to ourselves, and to another human being, the exact nature of our wrongs and the wrongs committed in the name of motorists or on behalf of the automobile-petroleum-highway system.

6. We were entirely ready to overcome all the obstacles—laziness or sedentary lifestyle, lack of equipment or infrastructure, lack of supportive friends or family, low-density development, misguided zoning laws or planning policies, unenlightened officials, industrial might, etc.—standing in the way of personal and social change.

7. We humbly asked Those In Power to remove any remaining obstacles, and when and where this failed, we took personal and community action to remove them (the obstacles as well as Those In Power).

8. We made a list of all persons and beings past, present and future, our driving had harmed, and became willing to make amends to them all.

9. We made direct amends to such people and beings wherever possible, except when to do so would injure them or others.

10. We continued to take inventory of our ongoing progress and impacts, and when our behaviour caused unnecessary damage to others or the environment, we promptly admitted it and took action to mitigate it.

11. We sought through spontaneous and convivial interaction to improve our contact with Neighbours and Community Members, reclaiming where possible the public space lost to the rush of traffic.

12. Having had a personal awakening as the result of these steps, we tried to carry this message to other autoholics, and to practice these principles in all our affairs.

None of this is to suggest that responsibility for our current mess—noise and pollution, death and injury, autoblight and urban sprawl—rests chiefly on the shoulders of the individual driver. On the contrary, the existence of a powerful and self-serving transportation industry has a lot to do with why alternatives have lagged so far behind. This isn't another conspiracy theory, it's simply a reason why the car has become in many cases the most convenient and practical option—as opposed to traditional car-free cities like Venice, Italy, where we obviously wouldn't need this program.

Just as in former Communist countries people were not hypocrites for opposing the regime while relying upon it to put food on the table, there is no contradiction in reluctantly using the car when and where it has genuinely become the only viable option. A regime is a system imposed from above—one does not always have the luxury of nonparticipation. Still, one's opposition to the regime is crucial, as acceptance precludes action, and even desire for change.

So the entry point for Autoholics Anonymous is a desire to take an individual stand, no matter how small and incomplete, against auto tyranny and above all, for a happy, healthy, independent life. But let's relax and take our time with this. There's no point in taking on a huge project and finding it so overwhelming that we quickly back out and forget about it. Better to break this down into realistic, manageable yet reasonably challenging steps. Let's forget about guilt, blame or notions of purity. Let's just go at our own pace and not focus too much on success or failure. We won't always live up to our goals, but that's to be expected and it's no reason to stop trying. So let's enjoy the positive benefits and have fun with it!

So, take the Car-Free Pledge yourself, see how you do in the two-week program. If you can manage a car-free week once, you might do it again. And again. And if you've already gone thought the detox process, why not hold seminars or workshops, helping others to overcome their problems? Pass around the pledges to addicts among friends and family.

The Car-Free Pledge

Congratulations on making this attempt to break your car dependency. Thanks for taking part.

This is intended to provide motivation and pointers to help you plan for the two weeks and succeed. You'll find day-by-day tips, suggesting ideas and steps you can take to make the most out of car-freedom, and should your will falter, you can work out how much money is going up in fumes by using the Count Up Your Car Costs form. Hopefully, you'll be mentally prepared, having read the 12-step Autoholics Anonymous program.

The golden rule: Plan ahead! Cars, mobile phones, emails everything conditions us to an instant lifestyle, everything on demand and at high speed. So let's slow down a bit and think ahead. Before you begin your period of abstinence, decide which combination of transport modes is going to replace your car. The bicycle, public transport, longer walks, taxis and car sharing might all play their part in getting us around. And another question we should be asking ourselves is: do we need to go on that trip at all? Can we cut down on the number of trips we do perhaps by using local shops and facilities, or by doing many tasks in one trip?

Another golden rule to remember: there are always alternatives to car-dependency.

A Few Good Reasons to Go Car-Free

These are just some of the things you might like to reflect on if your will weakens during the fifteen days you're trying to curb your habit.

Your Car Costs—how much goes into that machine?

It's a significant drain on your resources. On average in the US, maintaining a car takes between 10 and 20% of a motorist's total income. Fill in the figures below to work out how much you're saving:

Car Tax ____________ + Breakdown organisation subscription ______+ Insurance ___________ + MOT Test (UK only) fee __________ + Garage Costs (rent & tax) _____ + Estimated Annual Repair Bill __________ Gives year total: ____________ From the week before you start: Fuel Costs __________ + Parking costs ______________ + Tolls ___________ Gives week total: ___________

So your car-free weeks represent a potential saving of (year costs / 24 + 2 x week costs =) ______________________

Try to imagine that you had this much extra income. It's a good bonus, isn't it?

And Then There's The Rest of the World

An average motorist switching to public transport for two weeks will save the world three pounds (1.4kg) of atmospheric pollution. Switching to cycling or walking will save even more. Most cars emit over 1,000 pollutants, which cause acid air and rain, cancer, lead poisoning, a variety of bronchial and respiratory illnesses, brain damage in children, heart disease and global warming. In the US, between 10,000 and 24,000 people die each year as a result of pollution traceable to motor transport. Road transport is also the biggest single worldwide contributor to greenhouse gases, causing 30% of all CO_2 emissions worldwide.

There are many organisations that will do their utmost to help all your efforts to become car-free, for the two weeks and beyond. World Car-Free Day is co-ordinated internationally by Car Busters, an international transport activism collective based in Prague. Our website www.carbusters.org contains a searchable directory of groups all over the world working to reduce car usage by various means, which you may find useful.

Before the event. Prepare yourself well. If your job is genuinely out of your reach by public transport or bicycle, try to arrange someone to car-share with for the two weeks (and why not see if others at work will do the same?). Look up the public transport routes. Is there some kind of period ticket you can buy yourself for the duration? Are the times convenient? And if they really aren't, maybe you can even re-arrange your working hours around them, or work from home for some of the time. If you're planning to go by bike, get it out, give it some attention, buy a sturdy lock, lights and mudguards vital in anything less than perfect conditions. If it's a really good bike, consider getting it security marked: your local police will probably be able to do this. If you're riding on-road, slick tyres will make riding much more efficient. If you carry a lot into work each day, you might want to invest in some panniers, or a trailer or child seat if you're transporting small children. And if you haven't got a bike, hiring one for two weeks will probably cost less than your petrol costs, or you might be able to borrow one from a friend.

Make a count of all your car trips and costs in the week before you start trying to cut down, so you can see the results of your efforts. Only you can decide what the best way is for you, but if you're not at least partly equipped with a bike or knowledge of your public transport routes, you may well not make it. But if you do, maybe you'll do it again, and again ...

Day One: Friday. Start slowly today: Make the easiest journeys by public transport. If you're going out tonight, try biking, taking a taxi (find out these useful telephone numbers), or getting someone to drive you. If you go shopping, make an effort to think ahead, so you won't have to drive out in the next week to get something you forgot.

Days Two and Three: The Weekend. If you haven't done so yet, use this time to check the routes you'll be doing, by public transport, walking or cycling. There are probably more pleasant routes by bike or foot than simply following the car route. Stretch your legs a bit, go for a ride or a walk, make sure everything's in working order (on you and/or your bike).

Day Four: Monday. So can you make this your first car-free day? Try at least to make all your regular journeys without using the car, and if anything comes up unexpectedly, think: how would I deal with this without a car? What might make it easier?

Days Five to Seven: Tuesday–Thursday. If you've been feeling more tired than usual from all this moving about under your own steam, feel good about yourself. You're getting fit, and you'll adjust very soon. All through this week, try to keep a count of the times that you've absolutely had to use the car for something, and each time think how you might avoid it next time. If it's something your job demands on you, try to work with your employers to make things easier for all staff to go car-free.

Days Eight to Ten: Friday–Sunday. Make this a car-free weekend. If you do a big shopping trip, take along a sturdy rucksack or panniers. If you're shopping for a family, can someone else help you? Use the time and your experience of the last week to really prepare to make this week properly car-free.

Days Eleven to Fourteen: Monday–Thursday. Best of luck making this week totally car-free. This is what you took the pledge for. You'll probably be noticing by now how unpleasant cars make life for every other mode of transport. And hopefully by now you'll be feeling physically healthier.

Day Fifteen: Friday. Well, you've made it. Perhaps you might celebrate your achievement by participating in a Critical Mass, going on in a city near you (a good but incomplete list of rides is available at www.criticalmasshub.com). We hope you can use your experience in these weeks to cut down on your car use in the future.

Bibliography

Alvord, Katie. *Divorce your Car! Ending the love affair with the automobile.* New Society Publishers, 2000.

Avidor, Ken. *Roadkill Bill.* Car Busters, 2001.

Bradbury, Ray. *S is for Space.* Doubleday, 1966 / Bantam Books—Reissue edition 1970.

Crawford, J. H. *Carfree Cities.* International Books, 2000.

Ellison, Harlan. *The Essential Ellison: A 50-Year Retrospective.* Edited by Dowling, Terry, with Richard Delap and Gil Lamont, Morpheus International, 2001.

Kay, Jane Holtz. *Asphalt Nation: How the Automobile Took Over America and How We Can Take it Back.* Crown Publishers, Inc., 1997.

Naparstek, Aaron. *Honku: The Zen Antidote to Road Rage.* Villard Books, 2003.

Singer, Andy. *CARtoons.* Car Busters, 2001.

About the Authors and Artists

Wes Alderson. A writer of essays covering science, economics, and ethics. He is also the author of literary, science fiction and fantasy short stories — his first love. In 2001, shortly after his retirement from the electronics and acoustics industry, he earned a Certified Professional Writer degree from U.C.L.A Extension in Los Angeles, California, and he plans to spend the rest of his life writing.

Attila the Stockbroker. Has earned his living as a poet and musician for over 20 years, travelling all over the world. For everything you could ever need to know go to his website at www.attilathestockbroker.com.

Ken Avidor. Lives in a house on the south side of Minneapolis with his wife, two daughters, two cats and nearly all the modern technologies that make him crabby. He writes and draws "Roadkill Bill" for the Twin Cities Alternative Weekly Pulse. His comics, illustrations and articles have also appeared in Funny Times and Car Busters Magazine. Car Busters Press has published a collection of the early Roadkill Bill comics. Roadkill Bill is also featured in "Toward the Livable City", an anthology about urban design published by Milkweed Editions. Visit the the Roadkillbill.com website to learn more.

Greg Blanchette. Author, adventurer, engineer and free spirit. He grew up in Montreal, was based in Vancouver for 20 years, and recently fled to Ucluelet, a small town on Vancouver Island, where he works part-time in the newspaper business. Greg is a lifelong commuter and touring cyclist who decries the pervasive effect of the automobile on both society and nature. His work has appeared in many publications, from the *Globe and Mail* to *Sea Kayaker*.

Ray Bradbury. One of the most celebrated fiction writers of our time and the author of more than thirty books. Among his best known works are *Fahrenheit 451*, *The Martian Chronicles*, *The Illustrated Man*, *Dandelion Wine*, and *Something Wicked This Way Comes*. He has written for the theater and the cinema, including the screenplay for John Huston's classic film adaptation of *Moby Dick*, and was nominated for an Academy Award. He adapted sixty-five of his stories for television's *The Ray Bradbury Theater* and won an Emmy for his teleplay of *The Halloween Tree*. In 2000, Ray Bradbury was honored by the National Book Foundation with a medal for Distinguished Contribution to American Letters. Among his most recent books are *Let's All Kill Constance*, *From The Dust Returned*—selected as one of the Best Books of the Year by the Los Angeles Times—and *One More For The Road*. He lives in Los Angeles with his wife, Marguerite.

Michael Burton. A Chicago-based poet and bicycle activist. He founded the grass roots movement to depave Chicago's Lake Shore Drive and restore the super highway area as a carfree shoreline park. More information is at www.ForeverClearAndFree.org.

Sue Clancy. Looks at life through her funny-bone. She firmly believes that humor helps us look honestly and pragmatically at the world around us. It is through this frank examination and laughing at ourselves that we can begin to change things. Sue does humorous cartoons, illustrations and fine art works. She is published frequently in alternative newspapers like the *Oklahoma Gazette* and various "underground" Zines. She is the founder of *This Artist Studio* in Norman, Oklahoma, which creates hand-sewn books made with a sense of humor. Life's a funny beast, pet it gently! Website: http://home.telepath.com/˜artist; Email is artist@telepath.com.

Stewart Clayten. Has a BA in English from Simon Fraser University. He lived in Japan between 1992 and 1997, and currently resides in New Westminster, BC, Canada. He hopes to live in Japan again in the future. He is currently working on a novel in which the protagonist more or less stays inside his home.

Kevin Cook. Resides in Kitchener, Ontario where he runs his own business creating illustrations, graphic design, dvd designs, animation, photography and websites. He has been illustrating for over 22 years.

Kenneth De Crescenzo. Lives in Sacramento, California where he works for the California Department of Transportation. He grew up in New York City where he acquired his first bike at the age of thirteen. Ken's entire professional career has been in the transportation industry, starting as a bus driver, moving to transit system and highway planning, and then to his current position. He lives near the American River Bike Trail and frequently takes his blue Raleigh road bike out to see if he can make another "personal best".

Harlan Ellison. Has written or edited 75 books; more than 1700 stories, essays, articles, and newspaper columns; two dozen teleplays, and a dozen movies. He has won more awards for these works than any other living fantasist. He has drawn attention to the art of writing by performing the remarkable feat of creating and writing and completing stories in the windows of bookstores in full view of large audiences; stories that have won major awards and literary prizes. He lives with his wife, Susan, inside the Lost Aztec Temple of Mars, in Los Angeles.

Joanna Emery. Has been a freelance writer for over a decade. Her work has been published in many magazines on a variety of subjects ranging from cultural history to crop circles. Joanna especially enjoys writing for children: she has published one picture book, "Melville Smellville" and is currently working on two middle-grade books. She lives with her family in Dundas, Ontario and would like you to know that her husband and three children were the inspiration behind "Boiling Point".

Peter Gelman. Author of the satirical novels *Moonifest Destiny* (Quanta, 1995) and *Flying Saucers Over Hennepin* (Permeable Press, 1997). From Portland, Oregon, he writes for bicycle magazines. You can find more of his work in Breakaway Book's anthology *Bicycle Love* (2004). With Neal Skorpen, he's working on other projects. Email: petergelman@yahoo.com.

Robert Gregory. Has combined rich academic backgrounds at Cornell, Syracuse, and Duke Universities with a wide range of work experiences: vocational counseling, mental health research and practice, social science think tank membership, substance abuse program direction, anthropology in Vanuatu and elsewhere, and teaching and research about all aspects of rehabilitation and disability. His current passion is almost daily indulgence in tree planting on a 30 acre life-style block of land. Dr. Gregory teaches community psychology at Massey University in Palmerston North, New Zealand, and writes and publishes as much as possible about the many crucial matters influencing or about to change our little and lovely Planet Earth.

Matthew Frederick Davis Hemming. An animator, writer and amateur automotive ethologist based in the stinky megalopolis of Toronto, Canada. He has four cats, two dogs, one fish, one daughter and one wife. He has a shiny bicycle, and his favourite colour is Prussian Blue. Matthew can be contacted via e-mail at mfdh@mfdh.ca.

Lee Henderson. Vancouver's Fossilosopher; writes songs for community celebrations such as Bike Month, and The Dinosaurs Against Fossil Fuels Anarchist Street Theatre Collective.

Matt Hern. Lives in East Vancouver with his partner and daughters. He directs the Purple Thistle Centre, edits Crank Magazine and is on faculty at the Institute for Social Ecology. His new book is *Field Day: Getting Society out of School*.

Jim Hoehnle. A Vancouver based artist whose varied practice is focused on social change. Jim works in several different media including ceramics, painting, graffiti, and mixed media kinetic sculpture. One of Jim's most recent projects was presented at Vancouver's 2003 Parade of Lost Souls Festival. There, with the help of Pedal Works and local chopper bicycle enthusiasts, a chopper drama was performed, climaxing with the burning of an SUV in effigy. Children spontaneously cried out, "Burn the SUV", which was sweet music to the artists' ear.

Alex Lang. Poet, lyricist and television writer/producer; living in LA. He received his MFA from Antioch University, Los Angeles. His work has appeared in many periodicals, including *Caffeine*,

Recursive Angel, and *The Prospect (U.K.)*, as well as in the anthologies *The Brink: Postmodern Poetry 1965 To The Present*, and *The Road to Recovery: A Tribute to 9-11 and Other Poems* (as "Shuggie Love"). His lyrics are featured on the Monkey Bars' album *Food-Eating Food* (also as "Shuggie Love").

Janice Levy. Author of numerous children's books: "The Spirit of Tio Fernando," "Totally Uncool," "Abuelito Eats With His Fingers," "The Man Who Lived In A Hat," "Finding The Right Spot," and "Alley Oops!" She has won the Writer's Digest Magazine Competition for best literary short story three times.

Jeff Mann. A former 'back-to-the-lander' who has seen the error of this approach despite best intentions. He lives with his wife and two sons in a solar and wind powered, wood heated, compost toileted etc. house near the coast of Maine. He finds living there great, but regrets this life style is hopelessly dependent on the car. Aside from sculpture, he works as a substitute teacher and is a mediator for the Maine courts.

Scott Massey. A photo-based artist currently living in Vancouver. He recently completed a BFA (Photography) at Emily Carr Institute of Art+Design and has a background in furniture making and design. His work deals mainly with aspects of landscape that display evidence of the reciprocal relationship between nature and culture. Recently, he has been creating temporary, outdoor sculptural installations related to this theme.

Keith McKellar. Worked at sixty-five different jobs in several fields, before becoming the professional artist Laughing Hand. His travels have taken him on many urban vagabond walking journeys in North America and overseas. He has worked as a street artist for twenty-one years doing stints in California, Japan and Taiwan. His recent "Neon Eulogy; Vancouver Café and Street" is a collection of line-drawings and stories of Vancouver vintage cafes and cinema theatres that chronicles the rise and fall of neon in historical Vancouver and details an entertainment history of its cafes and theatres. He has published writings and drawings in a number of magazines and newspapers. See more at www.laughinghand.com

Ray McManus. Received his MFA in poetry from the University of South Carolina in May 2000. His poetry has appeared in *Crazy Horse*, *Nimrod*, *Oakland Review*, *Illuminations*, *Natural Bridge*, and other journals. He teaches poetry and composition at the University of South Carolina. Ray also coordinates and teaches community poetry workshops in South Carolina. "Gridlock","Pavement" and "Main Street at Eighty" are from "Asphalt", a series of poems about the road.

Rick Millikan. Has long enjoyed cycling as a practical "transsport" and life sport. A retired educator, he is now dedicated to sharing his enthusiasm for cycling in a monthly cycling column in Backpackers News, a British Columbia-Alberta publication. Moreover, Rick and his wife Chris encourage others to discover soft pedal adventures through their travel articles, especially in the beautiful bicycle friendly Northwest of British Columbia. Their travel stories have appeared in various newspapers as well as popular online magazines such as travel-wise.com, travellady.com and gonomad.com.

Scott Munn. A writer and photographer living in rural East Sussex, England. He is often faced with a traffic jam of several hundred sheep in his front yard.

Aaron Naparstek. A writer, interactive media producer and transportation activist in Brooklyn, New York. The never-ending assault of horn-blasting beneath his apartment window compelled him to begin writing honku—haiku poetry about honking, traffic and American automotive insanity. He began taping honku to neighborhood lampposts and soon a local traffic calming and poetry movement was born. Visit his web site, www.honku.org, and add your own honku to the virtual "Lamppost." His book, "Honku: The Zen Antidote to Road Rage" was published by Villard, June 2003.

Sara Ross. (a.k.a. RedSara) A locally engaged media producer in Vancouver, Canada. The focus of her work is creativity, community, and bicycles. A proclaimed cycling celebrationist and community artist, she brings art and shared creativity into our everyday experience of the city through projects such as the Democratic Poster Project, Community Public Art on bike routes, Critical Mass, Bikes Not Bombs Vancouver, and the Pedaling Zoo. She is currently working for PEDAL (Pedal Energy Development Alternatives) to promote bicycle culture, and to build community through bicycle creativity and bicycle art. Sara's professional art practice includes urban landscape painting, installations, graffiti, and performance art (online at www.SaraRoss.com). Her art education is ongoing, life-long, and mostly non-institutional.

Stephanie Scarborough. A pisces, vegan and wishes she could play the accordion. She writes a monthly zine called *The Cheap Vegan* and runs an online distro dedicated solely to food zines (http://diystore.cjb.net). Her work has appeared in *Whistling Shade, Haz Mat Review, Bovine Free Wyoming, Paperplates, Planet Relish, Tenderfoot* and elsewhere.

Andrea Schmidt. Lives in Vancouver, Canada with cat Bruce and roommate Lili. Currently works at the University of British Columbia designing books, posters, forms etc. Spends spare time biking aimlessly through the streets taking digital photographs. URL: www.a-schmidt.com.

Gene Seibel. A Kansas farm boy who left the country for the lights of the big city. After a few years in television broadcasting, he got back in touch with the great outdoors and wide open sky as a private pilot. Over the next twenty years, his love of flying took him, his family, and his flying machine into countless situations of adventure and intrigue all over this land of ours. Gene has always considered his opportunities to fly and travel a special blessing. He has shared them in any way he could, from giving airplane rides to children of all ages to writing about his experiences and observations. He is the author of the true adventure book *Confessions of a Pilot*.

Andy Singer. A car-free cartoonist and illustrator, currently living in St. Paul, Minnesota. His panel cartoons appear regularly in 17 publications, including The Utne Reader, The Funny Times, The St. Paul Pioneer Press, Ventura County Reporter, San Diego CityBeat, Boston's Weekly Dig and Z Magazine. He has also published cartoons in The New Yorker, Esquire, The Progressive, Discover and dozens of other magazines, newspapers and books around the world. He is the author of a book of essays and cartoons about automobiles entitled "CARtoons" and his work has appeared in two major cartoon compilations — "Attitude", edited by Ted Rall, and "The Best of the Best American Humor", edited by Raymond Lesser and Susan Wolpert of The Funny Times. You can see more of Andy's work at www.andysinger.com. Contact him at andy@andysinger.com

Neal Skorpen. Lives in Portland, Oregon with his wife Marcie and two cats. He is the creator of two comic strips *Cyclotoon*

and *Muddlemarch*, both of which appear in papers across America. Neal also teaches drawing and computer graphics at the Art Institute of Portland, practices Aikido, and plays too many video games. Visit his cartoon world at www.nealskorpen.com.

Michael Smith. Grew up in rural Kentucky and has lived for the last 25 years in New York. He has never owned a car. With Charles Komanoff and a few other other wild-eyed visionaries, he co-founded Right Of Way (www.rightofway.org) in the late Nineties; his *Manifesto* in this collection was originally written for the group's Web site. His work has appeared in the New York Times, the New York Daily News, Newsday, the Washington Post, and the online magazine Salon.

Karen I. Stanley. Can be seen around Vancouver riding an old five-speed now considered hip as a 'cruiser'. A yogini, community dancer and movement researcher, this loving Auntie prefers to walk and cycle so she can give attention to the detail surroundng us. Karen is a big fan of Dinosaurs Against Fossil Fuels and Chopper builder/riders; she rode a heavily modified trike as Good Witch for the Parade of Lost Souls this year. Her inner fire is stoked by: spiritual pursuits, weird or sometimes unacceptable art, cycling and a certain Ironworker.

Greg Taylor. A native of Alexandria, Virginia. When he isn't eating doughnuts or racing his bike for the Team Lardbutt Bicycle Club, he's usually drinking coffee. Identified by name in over one-third of ancient world's creation myths, Mr. Taylor is recognized by friends and family as a world-class smart aleck. He commutes to work in Washington, DC, via his bicycle. Mr. Taylor is also a semi-regular contributor to Cyclingnews.com.

Grace Tierney. A freelance writer who lives in Ireland and once upon a time she let a much loved bicycle rust in the back garden. When she's not writing, she enjoys gluttony, reading, gardening, and travel. Her short stories have been published in the US and UK, and her non-fiction has been published in Ireland, the UK, America, and Australia. She is a staff writer with 'Writer Online' and 'Netsurfer Digest'. More of Grace's work may be viewed at www.gracetierney.com.

Nicholas Wees. A native of Montreal; after travelling throughout North America he settled in Victoria, British Columbia, where he works in the printing industry. In addition to his love of literature, he is an amateur musician and dedicated gardener. He lives with his wife, who is a sculptor, his son and three cats.

Patricia Wellingham-Jones. Former psychology researcher and writer/editor, is a two-time Pushcart Prize nominee. She has work published in numerous anthologies, journals and Internet magazines. Joining the chapbook, *Don't Turn Away: Poems About Breast Cancer*, are her new collections, *Apple Blossoms at Eye Level* and *Bags*. *A Gathering Glance* is #43 in the Lummox Press Little Red Book series. Her winning poem was set to music in the SongText Contest, 2003 (www.songsforall.com). Her website is www.snowcrest.net/pamelaj/wellinghamjones/home.htm.

Dean Wirth. Author of many as yet to be published horror and/or satirical stories. His surreal courier comic "Overnight Sensations" is currently running in the *Momentum* magazine. Dean cites influences ranging from Lenny Bruce to Mary Shelly to Emily Dickenson to Theodore Sturgeon to John Wyndham to Woody Allen. He is currently working on a comic book version of "Overnight Sensations". Email is deanoramo@hotmail.com.

Jeff Younger. Prolific performer, composer and teacher; born in Winnipeg, Manitoba. Currently resides in Vancouver, British Columbia. He began playing guitar as a wee lad and has since contributed his particularly inspired creative energies to countless and madly varied projects. Straight up and experimental jazz, funk, folk, rock, live theatre and multi media collaborations and improvisations have filled his gigging calendar, while his compositional impulses have led him to write quirky jazz tunes, strange instrumental country music, funk extravaganzas, and sensitive folk songs. Currently, Jeff is finishing a bachelor's degree in composition, writing "serious" music in varying styles and instrumentations. His website is jeffyounger.com.

Organizations

Carbusters
Kratka 26
100 00 Praha 10
Czech Republic
Website: www.carbusters.ecn.cz

Right of Way
New York
USA
Website: www.rightofway.org

WALK Austin
P.O Box 7753
Austin TX
78713
USA
Website: www.io.com/˜snm/walk

Copyright Acknowledgements

About the Editor. Stephan Wehner was born in Fontainebleau, France. He grew up in Germany and Belgium, and now lives in Vancouver, Canada, with his partner Tamara Dakic and daughter Isabella. With a Master's of Science in computing science and a Ph.D. in mathematics he works as a software engineer and usually devotes most of his free time and mental energy to a math problem.

Ordering Information

Traffic Life: Passionate Tales and Exit Strategies may be purchased from the publisher directly by sending a cheque or credit card information for $21.90 plus $2.00 for shipping and handling plus GST (for Canada), $16.90 plus $3.00 (for the USA) to

wandering soliton publications
(a division of)
The Buckmaster Institute, Inc.
2150 Adanac Street
Vancouver BC V5L 2E7
Canada

Make cheques payable to *wandering soliton publications.*
Credit card orders may also be faxed to
(604) 648-8203
or made through the book's website:
www.trafficlife.com

Volume discounts are available on request.